The Colossus of 1812:
An American Engineering Superlative

by Lee H. Nelson

Published by the
American Society of Civil Engineers
345 East 47th Street
New York, New York 10017-2398

ABSTRACT

This book presents the developments in bridge building that occurred in the United States toward the end of the eighteenth century and in the beginning of the nineteenth century. With the growing need for bridges and the willingness to invest in such enterprises, American bridge builders were encouraged to be daring and inventive in their design. Their inventiveness led to the development of long-span wooden bridges with laminated members where the laminated members were used not only for major chords but for arched ribs in compression as well. This structural evolution in bridge design culminated with the building of the "Colossus" of Philadelphia, a 340 ft clear span wooden bridge designed and built by Lewis Wernwag in 1812. After explaining the historical context of this superstructure, the book then discusses "Colossus" in relationship to its wind bracing, abutments, and structural defects. In addition, a summary of a computer analysis of the bridge is presented. Due to Wernwag's innovated and superlative design, "Colossus" captured the imagination of both the romantic and technological minds of the day and influenced American bridge building for some time to come.

Library of Congress Cataloging-in-Publication Data

Nelson, Lee H.
   The colossus of 1812: an American engineering superlative/by Lee H. Nelson.
      p.   cm.
   ISBN 0-87262-737-9
   1. Lancaster-Schuylkill Bridge (Philadelphia, Pa.) 2. Bridges, Wooden—Pennsylvania—Philadelphia—Design and construction—History. 3. Philadelphia (Pa.)—Buildings, structures, etc.
I. Title.
TG25.P52N45   1990
624'.6'0974811—dc20   89-18612
                                                                CIP

Preface

Mr. Nelson deserves the thanks of all civil engineers for his scholarship and dedication in producing this oustanding book. It concerns one of the greatest structures ever produced by an American civil engineer, but one that has received very little attention by historians. To put this in perspective, "The Colossus" had the greatest span [340 ft] of any wooden bridge ever built. [N.B. Burr's McCall's Ferry Bridge, which lasted only a very short time, may have had a slightly greater span, but this has never been verified.] In any case, with the publication of this book a long-term historical oversight will have been rectified.

The writing of this book is a story in itself. Mr. Nelson's interest in bridges stems from his boyhood in Portland Oregon, which is famous for its many crossings of the Willmette River. It was largely due to his publication of 'Oregon Covered Bridges,' in about 1960, that a citizen's effort to save the state's remaining structures was inaugurated. Later, Mr. Nelson went to Philadelphia to work on the preservation of Independence Hall and other historic structures. In the library of the Historical Society of Pennsylvania he came across the papers of the Colossus Bridge. Then, for over 15 years he pursued the task of writing this book.

Mr. Nelson, an architect, has spent his profession-al career in the rewarding field of historic preservation and so has an intimate knowledge of structures. He couples this knowledge with his sense of history to produce a book that gives the reader both technical and historical insights which give it a double appeal to both civil engineers and historians.

Although it has been the long-term policy of the American Society of Civil Engineers to publish outstanding scholarly works on civil engineering history, there have been far too few to have even been considered. When this book was received by the Committee on the History and Heritage of American Civil Engineering, it was unanimously approved for publication.

It is hoped that the publication of this book will encourage architects, civil engineers, and historians to undertake the necessary research and writing to produce other works of this nature and caliber. Manuscripts or even ideas would be welcomed by the Committee.

Neal FitzSimons, Fellow ASCE
Chairman, Committee on
History and Heritage of
American Civil Engineering

CONTENTS

ILLUSTRATIONS

Introduction

The last decade of the eighteenth century and the first decade of the nineteenth century was one of those incredibly innovative periods in the history of American bridging—a period that encompassed the building of mighty wooden bridges in America—bridges that were daring (or reckless) by modern standards—with builders that were inventive in their efforts to eliminate the traditional (and inefficient) mortise and tenon joints—structures that pushed American wooden bridge technology a bit beyond the state of the art as it then existed in Europe—and, bridges that met the almost unrealistic expectations of local governments and speculators for the economic expansion of America in the early days of the Republic.

When you look at the European bridge construction context (primarily German and Swiss) for the latter part of the eighteenth century as it might have influenced American bridge builders, you are left with the impression that American builders did surprisingly little borrowing from the Old World for their structural ideas, despite the fact that some European wooden bridges were well known in America through builder's treatises, traveler's accounts, and published engravings of the bridges. From these early views and from the rare surviving wooden bridges in Switzerland, we can draw some tentative conclusions about American bridges at the end of the eighteenth century. First, they had a *structural design clarity* that was lacking in most European wooden bridges, which were then burdened with very complicated, wood-wasteful carpentry, involving multiple Queen-post trusses within a given bridge, resulting in many diagonal members between the vertical posts, and with the most dazzling, labor-intensive use of continuous zig-zag joints, secured with wedges, to laminate several timbers into a larger chord member. American bridges tended to have structural bays that, by present standards, were more rational, that is, consisting of simple triangles rather than structural bays with polygonal shapes.

There had been a long tradition of wooden truss bridges in Germany and Switzerland, where (like America) there was a plentiful supply of strong, straight timbers, and some of their eighteenth century bridges were well known to Americans. The bridge at Schaffhausen, in Switzerland, for example, was famous in its own time (from engravings and published descriptions); but neither the Schaffhausen bridge nor its contemporaries seem to have had much *structural* influence in America, though they did demonstrate that long-span bridges could be built of timber (Figs. 1–3).

Starting in the last decade of the eighteenth century, American bridge builders began to demonstrate their own daring and inventiveness in the development of long-span wooden bridges with laminated members; and, shortly after the turn of the nineteenth century, some American bridges utilized laminated members not only for major chords, but for *arched ribs* in compression as well. Perhaps the use of such arched ribs is the most interesting aspect of this early American "track" of laminated bridge construction.

Today, of course, there are more modern definitions for the term *laminated* wood, but, with our historic focus on laminated arch construction, we are defining it as meaning wooden arches consisting of *multiple members* of wood, bound together with bolts or pins or iron bands so as to intentionally act as one larger composite arched structural member, bearing against the *faces* of the abutments rather than resting on *top* of the abutments.

The leading American theoreticians and experimenters in this new phase of laminated construction were Charles Willson Peale (1741–1827), whose contributions to the subject have (unfortunately) been largely discounted; Timothy Palmer (1751–1821), a productive and innovative Yankee bridge builder whose long-lived constructions began to "bridge" the gap between conventional carpentry and more efficient ways of spanning American waterways; and Theodore Burr (1771–1822), an ingenious and daring (and perhaps overly ambitious) builder whose designs attracted European interest and had a long-lasting impact on American bridge building.

The remarkable structural evolution that marked the latter stages of American engineering empiricism in the early nineteenth century was culminated in a dramatic way with the "Colossus" of Philadelphia, a 340-ft clear span wooden bridge designed and built (and named) by Lewis Wernwag in 1812. It was world famous in its own time, and for good reason. By Victorian standards, it was a romantically handsome structure in a picturesque setting adjacent to the Philadelphia Waterworks. It was the longest clear span wooden bridge in the world when it was built. For that reason, as well as for its setting, it was admired by engineers and artists from the United States and abroad. In short, it was America's premier engineering superlative of the early nineteenth century. In historical perspective, it was a world-class structure.

1

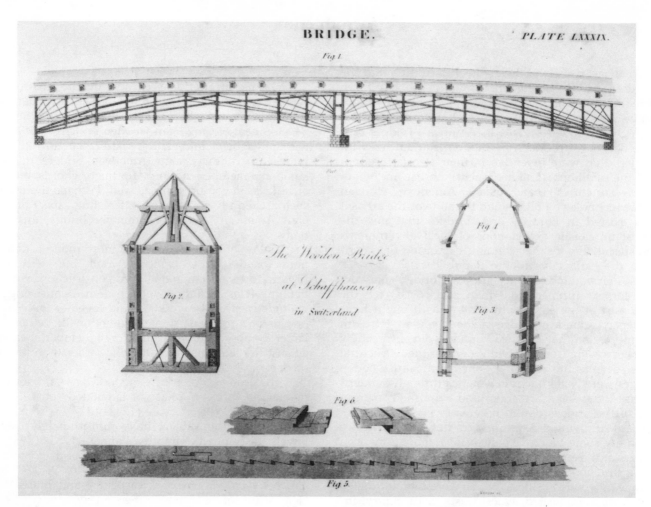

Fig 1

The Wooden Bridge
at Schaffhausen
in Switzerland

Fig 1

Fig 2

Fig 3

Fig 6

Fig 5

FIG. 1.

"The Wooden Bridge at Schaffhausen in Switzerland."
This bridge was internationally famous in its own time,
having been built 1756–1758 by Hans Ulrich Grubenmann of the then-celebrated Grubenmann family of
carpenters. He first proposed a single-span bridge, but
this was too daring for the authorities, who required
that the center pier from a previous multispan bridge
be reutilized. The story goes that Grubenmann cleverly
redesigned his single-span bridge into what only
appeared to be a two-span bridge (to satisfy the
authorities), but that it did not rest on the center pier
(to prove that he could do what he had proposed).
Despite this story, the consensus of opinion of those
who inspected the bridge was that it was a true two-
span bridge, and that it did rest on the center pier.
There are a number of discrepancies about its length
and its details, but the two unequal spans were
approximately 193 and 171 ft. It was a daring and
large-scale project, consuming 400 large fir trees. It
was visited and admired and depicted by many.

Although this bridge was atypical in terms of its
length and scale, it was similar to many other bridges
of the day in that structurally it was a multiple queen-
post system. The many resulting diagonals, with very
complicated roof framing, are what typifies many of the
Swiss bridges from the last half of the eighteenth
century. They were made even more complex with
their auxilary struts and longitudinal zig-zag scarf
joints, which laminated several timbers into one larger
chord. Fig. 5, along the bottom of this plate, shows the
zig-zag joint with its wedges that interlocked the
multiple members into one larger composite member.
Figure 6 (just above) shows a somewhat more conventional scarfed joint that spliced the ends of individual
timbers into longer continuous members. These aspects were typical of many less ambitious spans in
Germany and Switzerland.

Although it must be admitted that structural
details are not available on American bridges until the
latter 1790s, there is no evidence that American
bridges utilized such multiple queen-post trusses, nor
do they seem to have used other aspects of those
complicated carpentry systems, such as the zig-zag
continuous scarfing for laminating timbers.

Plate LXXXIX, entitled "BRIDGE," engraved by
[William] Kneass, a Philadelphia engraver, is probably
from one of the numerous encyclopedia published in
Philadelphia in the early nineteenth century. Courtesy
of Nicolas F. Veloz, Jr.

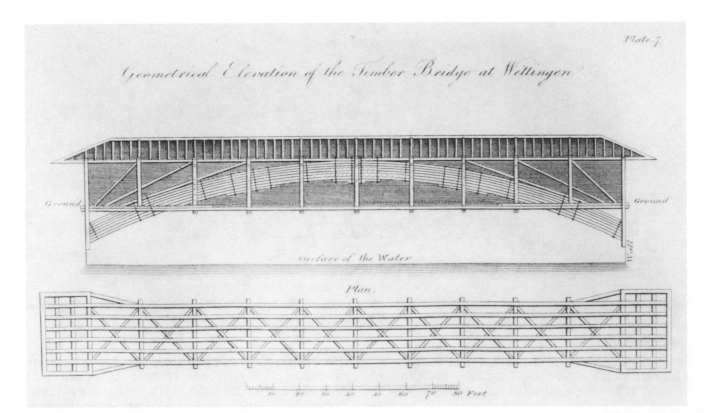

Plate 7

Geometrical Elevation of the Timber Bridge at Wettingen

Ground   Ground

Surface of the Water   Wall

Plan.

10   20   30   40   50   60   70   80 Feet

FIG. 2.

"Geometrical Elevation of the Timber Bridge at Wettingen." This remarkable structure, over the Limmat (near Baden in northeast Switzerland), was built in the mid-1760s by members of the Grubenmann family. Like the Schaffhausen bridge, the Wettingen bridge was well known and was visited by a number of engineers, architects, and interested travelers, resulting in a variety of written descriptions, dimensions, and details, none of which seem to have agreed except as to its general configuration. Its span was approximately 200 ft, with wooden frames supported by laminated arched ribs that were comprised of approximately seven courses of wood that were notched and banded together with iron hoops and keys, for a total rib depth of about seven feet.

If any European bridge can be said to have influenced an American bridge, it was probably the Wettingen Bridge. To the extent that it utilized laminated arched ribs that supported a level roadway, it was similar to, and may have influenced Theodore Burr in his construction of, the multispan bridge over the Delaware at Trenton in 1804–1806 (see Figs. 12–13).

This American view of the Wettingen Bridge is from the book by Thomas Pope ("Architect and Landscape Gardener"), *A Treatise on Bridge Architecture. . .* (New York, 1811), Plate 7. Courtesy of Eric DeLony.

3

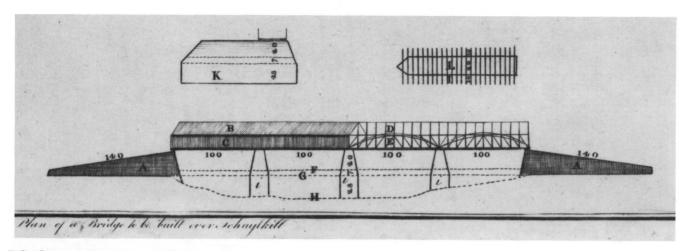

Plan of a Bridge to be built over Schuylkill

FIG. 3.

The engraving of this anonymous bridge design was published in the *Columbian Magazine,* Philadelphia, January, 1787. Although never built, this design is of interest to American bridge historians for several reasons.

First, though any scholar will agree that it is probably dangerous to assert that anything is an historical ''first,'' nevertheless, the engraving of this anonymous design may well be the first published design for a proposed American trussed bridge, that is, a design that is unquestionably a through truss (in this case, reinforced with an arch), as opposed to simple pile-and-beam or braced beam bridges.

Second, this design represents a later stage in the long-standing efforts to secure public funds to bridge the deep, tide water crossing of the Schuylkill River at Market Street in Philadelphia and thus to provide a reliable linkage with the country to the west. Though Philadelphia was served by ferry crossings, there was a long-felt need for all-weather crossings (secure against floods and ice) over the Schuylkill. This need had been articulated starting in 1750, when Benjamin Franklin was appointed (along with other responsible citizens, including Edmund Woolley, one of the builders of Independence Hall) to serve on a fact-finding committee for the General Assembly of the Province of

Pennsylvania, regarding the feasibility for building such a bridge. Nothing came of that, but there were numerous other proposals in later years. In the 1760s, for example, there were several proposals, including one for a 400-ft arch and another by the Philadelphia master builder/architect Robert Smith, who presented a plan, elevation, and model, the description of which bears a striking similarity to the engraving illustrated here, though Smith had died a decade before this engraving was published in 1787. Ultimately, the reason most of these proposals failed was the formidable challenge in building piers in such deep water.

This bridge design is also of interest because it suggests a very different approach to bridge design from any of the European bridges, especially the more typical ones with incredibly complicated carpentry (Fig. 1). This design consists of a simple mathematically rational truss with one diagonal brace in each bay, but with the truss being assisted by an arch on each side. It was very similar to the so-called ''Burr Truss'' (not illustrated here), which did not make its debut until nearly 20 years later. Perhaps Theodore Burr got the idea for an arch reinforced truss from this engraving.

Reproduced from the *Columbian Magazine,* January, 1787, facing p. 244, courtesy of the Library of the American Philosophical Society.

American Experiments with *Laminated* and Multiple-Membered Arches

From the mid-eighteenth century on, the bridging of America's coastal waterways to better connect cities with rural areas and thus to help promote the growth of commerce, agriculture, and manufacturing became a matter of considerable economic and political interest, an interest that was evidenced by many proposals to erect so-called "permanent" bridges (as opposed to the "unreliable" ferries) for the "greater conveniency" of local inhabitants. There were numerous mentions in newspapers, magazines, and other publications dealing with the need for bridges.

The public interest in proposed bridges was further evidenced in drawings and models of bridges that were exhibited in various cities to promote bridging ideas and proposals; and there were laudatory comments that extolled the virtues of bridges that had actually been built up and down the American seaboard.

By the end of the eighteenth century, a number of trestle bridges and modest trussed bridges had been built (including several impressive long-span bridges in New England); but several of the larger seaport cities, such as Philadelphia, would require very substantial bridges to cross the wide coastal rivers, involving costly and dangerous piers to be built in deep water. Even if they could be financed and accomplished, such structures would still pose an obstacle to navigation and intracoastal shipping. From the late 1780s through the end of the century, there was a serious interest in long-span trussed bridges and long-span arched bridges, as was evidenced by various proposals advanced by "ingenious" builders, self-taught architects, and others. The interest in long-span bridging efforts was being encouraged by the press and by investors alike, to say nothing of public and civic interest. The time was right for truly daring proposals. Little is known about most of these proposals; but one proposed long-span bridge design (uncovered), over the Schuylkill in Philadelphia, survives in a water-colored elevation by the French engineer Godofroi Du Jareau, dated 1796.[1]

The first American to design, promote, *and* patent a laminated arch bridge was Charles Willson Peale. In 1797 he published a 16-page pamphlet, which is interesting for several reasons. Nevertheless, the arch that he proposed was so fantastic that it has not been taken seriously by later scholars.[2] It would be difficult to assess its influence upon bridge building at the turn of the century, despite the fact that the French Academy of Sciences considered its merits (and problems); and while they felt that Delorme's system was probably preferable because it used less wood, that august and expert group certainly did not find any serious problems with Peale's ideas.[3] If anything, Peale's illustrated publication may have had some impact upon American bridge building, but that has

not yet been ascertained (see Fig. 4).

If Peale's design had a "fatal flaw," it probably was because the rise of his proposed arch was so great as to almost preclude the passage of heavily loaded wagons drawn by beasts of burden (see Fig. 5). Despite this flaw, the design is interesting as a constructional concept, involving modestly sized members (2 in. $\times$ 12 in.) treenailed together with locust pins to form an arched deck of six thicknesses for the full width of the bridge (some 50 ft), with five sets of prestressed "railings" (actually trusses) that would probably defy structural analysis (see Figs. 6 and 7). However, the laminated arch itself was conservatively designed as to the bearing area of the wooden arch in relation to the estimated live and dead loads. Even though the arch could have resisted the thrust, the bridge probably would have been excessively "springy," that is, lacking stiffness. Furthermore, such a bridge would have been extremely vulnerable to the elements. We do not know when the first *covered* bridge was constructed (despite claims from several contenders); and though Peale stated (in 1797) that some covered bridges had already been constructed in America, he considered doing so an "unnecessary expense," except to preserve the traditional mortise and tenon work commonly employed in trusswork. Peale would have coated his arch with a tar-based mixture (pity the horses).

While we have no evidence of any bridge having been built on Peale's plan, he did build a very large scale model that he exhibited in his museum in Philadelphia. With his publication, the model, the patent, and his connections in Philadelphia, we should not underestimate the possible impact of Peale's ideas. We cannot help being impressed with his design for a daring bridge, wherein he designed a true laminated arch span that studiously avoided the use of mortise and tenon joints. Peale laid the intellectual groundwork for his successors, namely, Palmer, Burr, and Wernwag, who built bridges utilizing multiple members on a grand scale—and Peale did this in 1797!

Although it was not built along Peale's designs, private financiers finally succeeded in constructing an impressive (and successful) "Permanent" Bridge in Philadelphia, which was built in 1801–1805, across the river Schuylkill at High Street (see Fig. 8). Incidentally, Benjamin Franklin had played a much earlier official (but unsuccessful) role in planning a bridge at that site as far back as the 1750s. The "Permanent" Bridge involved the best engineering talent available, namely that of William Weston, the English hydraulic engineer, together with Timothy Palmer, the master builder, and Owen Biddle, the architect of the covering.[4]

The success of this bridge triggered an incredible spate of other important bridges, especially in Penn-

cont'd on page 11

5

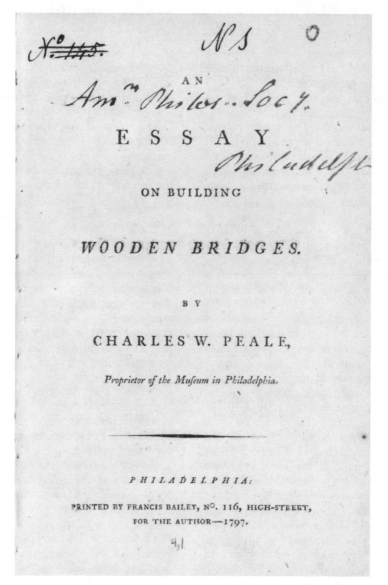

FIG. 4.
Title page from *An Essay on Building Wooden Bridges*, Charles W. Peale, (Philadelphia, 1797). Charles Willson Peale was the first American to design, promote, *and* patent a laminated wooden arch bridge. His patent was dated January 21, 1797, and it was the first bridge patent of any design in this country. His 16-page pamphlet (with six plates) was dated March 17, 1797, and it encompassed the general concept for the design and construction of a laminated arch (though he did not use that term), but his plates and descriptions were for a fantastic (technically feasible but probably impractical) bridge across the Schuylkill River with a clear span of 390 ft between the abutments!

He made a model of such a bridge, the arch of which was composed of four boards, 1 in. thick and 12 ft wide with an 80-ft chord, which he claimed would bear "more than" 200,000 pounds, which was "twice the weight of as many men as could possibly croud

[sic] together on such a foot Bridge." Peale offered to sell the patent rights to build a bridge from his invention, at the rate of two dollars for every 10 feet of span.

Perhaps because it was so unorthodox compared to the then-conventional truss bridges, few historians have taken Peale's design seriously. However, his 16-page pamphlet may have had some influence on other bridge builders, but that possibility has not been established. Although Theodore Burr, in 1804, appears to have built America's first true laminated arch bridge, which had some notable improvements over Peale's design, his bridge (at Trenton, see Figs. 12–13) has more in common with the Wettingen Bridge (Fig. 2) than it does with Peale's design. He could hardly have copied Peale anyway because of the patent laws. At that time, Switzerland did not have patent laws.

Illustration courtesy of the Library of the American Philosophical Society, Philadelphia.

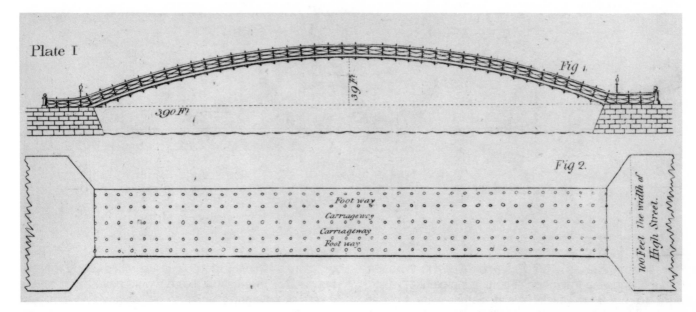

FIG. 5.

Plate I from Peale's 1797 pamphlet, showing his plan and elevation proposal for a 390-ft laminated wooden arch bridge over the Schuylkill at High Street (now Market) in Philadelphia. The width of the bridge would have been 50 ft, with a rise of 39 ft (probably its most impractical aspect); with five "ranges" of trussed railings (see Fig. 7) that divided the bridge into two footways of 8½ ft each and two carriageways of 12 ft each, with the remaining 9 ft of width taken up by the outer projections and by the stanchions of the railings.

Illustration courtesy of the American Philosophical Society, Philadelphia.

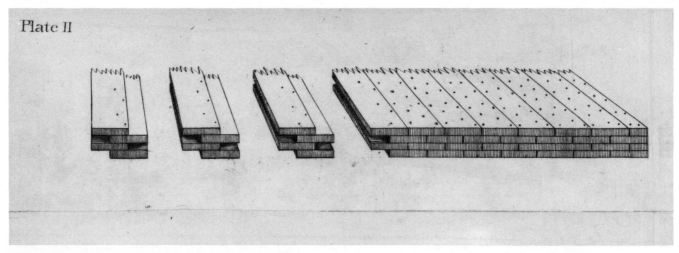

Plate II

FIG. 6.
Plate II from Peale's 1797 *Essay on Building Wooden Bridges,* showing the mode of laminating planks, which were to be 2 in. thick held together with Locust "trunnels" (treenails). Though this plate shows but four thicknesses, the Schuylkill bridge would have had six laminations, which would have been "placed butt to butt breaking joints, and firmly secured by keyed bolts at the butts of each of the outer planks, with a considerable number of trunnels, drove in like manner as is the practice in ship building; such a combination of planks will in fact be as a solid body of timber. . . ." In addition, there would have been several layers of covering planks, to be replaced as they "are liable to be worn out."

Peale further claimed an advantage of this mode of construction in that "the grain of all parts of the body of the Bridge being parallel, should the timber shrink or swell, no injury will happen to the work."

Illustration courtesy of the American Philosophical Society, Philadelphia.

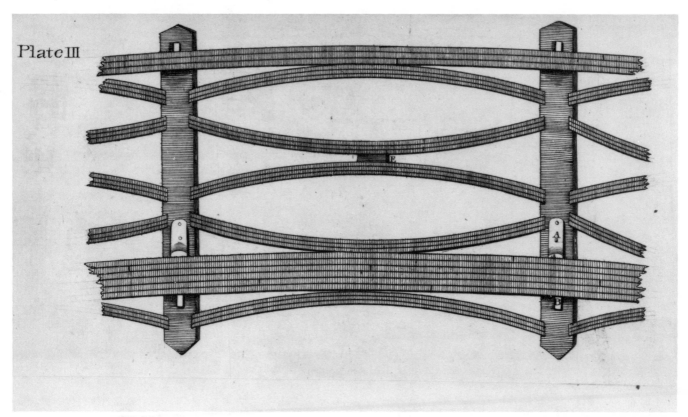

FIG. 7.
Plate No. III from Peale's 1797 pamphlet, showing a typical "railing," which was actually the means of "trussing" the laminated arch. Visible here are the six laminations of the arch, together with the "stanchions" of the railings and the curved (actually prestressed) members that comprised the "trussing" of the laminated arch. The stanchions were secured to the arch with ships knees and wedges (also see Fig. 14). Such ships knees were of course common to ship construction, but they were also used for building construction (witness the Independence Hall steeple reconstruction of 1828). Peale designed the outer railings to be 5 ft high, while the inner railings that were to divide the carriageways from the footways were to have been 7 ft high.

Illustration courtesy of the American Philosophical Society, Philadelphia.

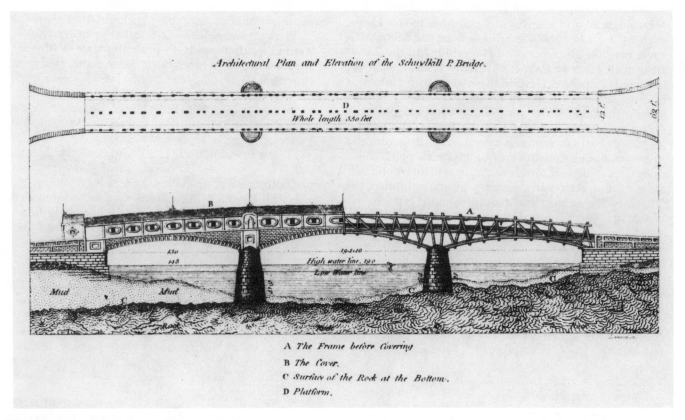

Architectural Plan and Elevation of the Schuylkill P. Bridge.

Whole length 550 feet

A *The Frame before Covering*
B *The Cover.*
C *Surface of the Rock at the Bottom.*
D *Platform.*

FIG. 8.

This is the "official" view of the "Schuylkill Permanent Bridge," Philadelphia, built 1801–1804 over the Schuylkill River at High Street (now Market Street, also the site of Charles Willson Peale's proposed design), the superstructure of which was designed and built by Timothy Palmer, and which was one of several early bridges built by Palmer utilizing arched ribs with multiple members. The Schuylkill bridge had three arches with a clear center span of 195 ft, and the two side arches were 150 ft in the clear. The 195-ft center span had a rise of 12 ft.

By virtue of the "laminated" ribs and the large member sizes, Palmer's bridges literally carried American bridge construction to new lengths. While this bridge was better known for the engineering involved with the construction of cofferdams for the piers (designed by the English engineer William Weston), another interesting aspect of the bridge was the use of multiple members for the ribs that were bolted together to form a unified arch. Unfortunately, there were no details of the trussing in the original report.

Among Palmer's other important bridges were the 244-ft arched span (about which little is known) over the Piscataqua, near Portsmouth, New Hampshire, in the mid-1790s and the bridge over the Delaware River at Easton, Pennsylvania, built 1805, which had three spans of 163 ft each. Like the "Permanent" Bridge, this latter structure had large-size members, "laminat-ed" with bolts to form the primary arch, but the Easton Bridge was improved over the "Permanent" Bridge in that it had a flat roadway (for an illustration, see Theodore Cooper's article in *Transactions,* of the American Society of Civil Engineers (ASCE), XXI, no. 418, July, 1889, plate VII).

This was the American state of the art at the turn of the nineteenth century, for it marked the beginning of the use of multiple members for arched ribs, and it lessened the traditional use of mortise and tenon joints with their susceptibility to dry rot; but it was to be followed by more innovative solutions, such as those used by Theodore Burr and Lewis Wernwag in the next few years.

The "Permanent" Bridge, opened for service in 1805, had a long life. It was altered and enlarged for the City Railroad c 1850, and it burned on November 20, 1875.

This plan and elevation, engraved by Alexander Lawson, was part of the "Statistical Account of the Schuylkill Permanent Bridge, Communicated to the Philadelphia Society of Agriculture, 1806," printed by Jane Aitken, Philadelphia, 1807, and published as part of the *Memoirs of the Philadelphia Society for Promoting Agriculture. . .,* vol. I (Philadelphia, 1808).

Illustration from the Lawson engraving, courtesy of the Smithsonian Institution, National Museum of American History.

cont'd from page 5

sylvania. This is not to overlook the fact that there had already been numerous and impressive bridges built in New England and elsewhere, beginning in the 1790s, such as, for example, the bridges over the Merrimack River near Newburyport and Haverhill, and the 244-ft clear span (also by Palmer) over the Piscataqua River in New Hampshire. Despite these early and ambitious efforts, America's bridging needs were largely unmet, mostly because of the formidable physical obstacles posed by eastern seaboard rivers, harbors, and estuaries; these barriers required very costly, difficult, and time-consuming construction of masonry piers, abutments, pile driving, and caissons. Furthermore, the state of the art regarding truss building limited the length of the spans, thus requiring more piers in the rivers, an especially difficult undertaking in tidewater areas.

The "Permanent" Bridge in Philadelphia was most likely the earliest major step in the direction of American multiple-membered, or *laminated*, arch construction. For the purposes of this book, we define *laminated* arch construction to mean those designs that utilized multiple-membered arched ribs acting in compression, which by virtue of their construction technology were attempting to function as larger composite structural members.

Although the three-arch "Permanent" Bridge over the Schuylkill at Market Street in Philadelphia has been frequently described as consisting of arched *trusses*, that is, trusses with a considerable camber, we will term them trussed *arches*, because the primary load-bearing members appear to have been the arched ribs, which (from contemporary drawings) were bearing *against* the piers and the abutments instead of resting *upon* them (see Fig. 9). Thus, the "Permanent" Bridge would qualify under this definition because it utilized double members, each measuring 14 in. by 16 in., bolted together creating three arched ribs in each of the three spans (the middle arch was a clear span of 194 ft, 10 in.). Thus, each of the composite arched ribs measured 16 in. by 28 in., and they embraced the posts by being cut away to receive the posts (see Figs. 10 and 11).[5] While the cutting away of those ribs to receive the posts may be regarded as a structural inefficiency, it was a step away from the more traditional mortise and tenon joints, and it led to the more efficient designs by Theodore Burr and Lewis Wernwag.

In its own day, this bridge cost an incredible sum, exceeding $250,000 of private speculative dollars, optimistically based upon the potential revenue from tolls (the annual average of which was $15,000 between 1805 and 1822). Furthermore, it required over three years of construction.[6] Whether that bridge might have been more daring is hard to say, but it seems unlikely considering the completion report, which makes the statement that, despite the desirability of a clear span for navigable rivers, the opinions of "practical men here" declared that a bridge of 200 ft "begins to be critical and hazardous."[7]

Nevertheless, there is no doubt that there were builders and engineers (self-styled or otherwise) who were more than ready to push the state of the art. It appears that the ideas for pushing the frontiers of structural "art" involved two relatively new ideas—the obviation of traditional trusswork in favor of arches or "ribs," and the idea of constructing those arched ribs with multiple members working in unison by being pegged or bolted together.

Prior to Palmer's 1805 "Permanent" Bridge, there had been several European efforts to build with multiple-membered arches, and at least one of those was well known to Americans who were interested in such matters; that was the Swiss bridge at Wettingen, built in the 1760s by the Grubenmann brothers, with a span of approximately 200 ft. Even better known was the Delorme method of building arches with multiple members aligned *vertically*; but it is hard to see that the Swiss experience or the Delorme system had any impact on American bridge building, though the latter's importance to dome construction has only recently been studied and appreciated in this country.[8] Other possible influences upon American laminated arch construction, such as the innovative multiple-membered arched bridges in Bavaria by Carl Friedrich von Wiebeking, were yet to be built.

Although the "Permanent" Bridge in Philadelphia marked an early step in the direction of American multiple-membered arched ribs, it was not "laminated" in the sense that Peale had proposed.

In 1804, the first great bridge utilizing true laminated arched ribs was built across the Delaware River in New Jersey, designed and built by Theodore Burr, who had built and would continue to build a succession of important and innovative bridges (see Figs. 12–13). He later built what may have been the longest single arch span wooden bridge ever built—the McCall's Ferry bridge over the Susquehanna River, reported to have had a clear span of 360 ft, but which was taken out by ice in March, 1818 after a very short useful "life." Little is known about its design.[9]

However, Burr's Trenton bridge must be regarded as a significant construction, partly because it was a precedent-setting design with a true laminated arch, not in the Delorme manner with vertical laminations, but with 4 in. by 12 in. members bent to the curve and laminated horizontally (like Peale, except that Burr's bridge had five separate ribs rather than a full-width laminated deck). Furthermore, it had a *flat* deck that was suspended with iron "chains" (actually square iron bars forged into hangers). The deck was "stayed" with diagonals to reduce the susceptibility of wave action in a suspended roadway. This bridge was widely admired, described, and illustrated both in the United States and abroad.[10] Its importance and innovative qualities should not be underestimated, and it

11

deserves its own study; but that is not within the scope of this book. Instead, it is our intent to focus on Burr's equally innovative, but perhaps more significant, rival Lewis Wernwag, who pushed the technology in new and interesting ways.

To summarize the early nineteenth century American contributions to multiple-membered arched construction, it is difficult (based on the available drawings and descriptions) to be sure of the structural behavior of these bridges; nevertheless, it is this writer's contention that Peale's patented design for a true laminated arched deck (perhaps impractical), but, more importantly, Palmer's "Permanent" Bridge in Philadelphia, Burr's very ingenious Trenton bridge, and, finally, Wernwag's "Colossus" (still to be discussed)—all comprise America's apparently separate "track" in the development of multiple-membered or "laminated" arch construction (for a structural comparison of these four bridges, see Fig. 14).

It is interesting that, considering the timber-poor situation in Britain at this point in history, there was a special interest in the three executed American wooden bridges previously mentioned, which resulted in their publication in various British engineering treatises and prints in the 1820s, 1830s, and 1840s.[11]

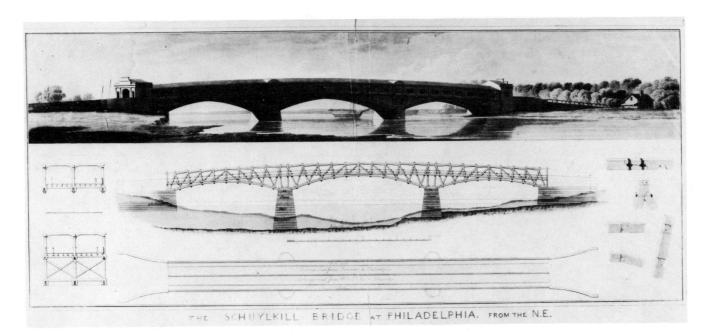

THE SCHUYLKILL BRIDGE AT PHILADELPHIA. FROM THE N.E.

FIG. 9.
While the "official" engraving by Lawson (see Fig. 8) showed the general configuration of the "Permanent" Bridge by Timothy Palmer, neither the Lawson view nor the official report gave any details of the wooden superstructure. Interestingly, the first view to give any such details was a drawing made by C.A. Busby (aquatint by M. Dubourg), which was published by J. Taylor, London, 1823.

The Busby drawing includes details and structural information not in the official report. It is not clear whether this represents a firsthand "inspection," but the "new" information consists of the two cross sections and some details that include a scarf joint for the upper chord and a mortise for a post, a post with its shouldered (joggled) connection for a diagonal brace, together with those details showing the half-lapped connections between the posts and the roadway arch and the primary bottom arch, the latter shown for the first time as a composite or "laminated" member.

This Busby/Dubourg aquatint is entitled *The Schuylkill Bridge at Philadelphia, from the N.E.* Courtesy of the Smithsonian Institution, National Museum of American History.

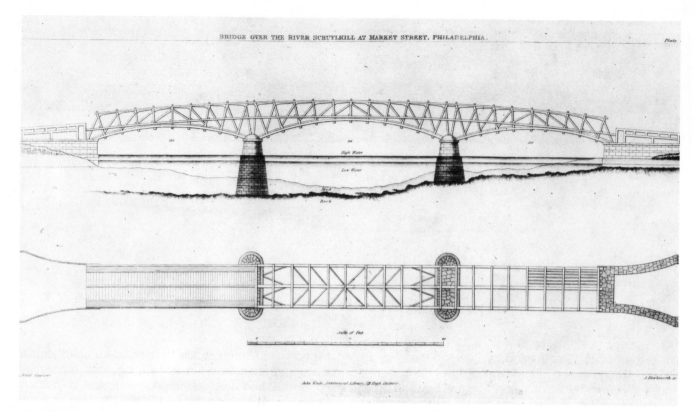

FIG. 10.
Another English view of the "Permanent" Bridge, entitled "Bridge Over the River Schuylkill at Market Street, Philadelphia," drawn by E. H. Gill, Engineer, and engraved by J. Hawksworth, was published as Plate 34 in John Weale's, *Bridges in Theory, Practice and Architecture* (London, 1839–1847).

The elevation drawing seen here appears to be a rather literal redrafting of the Lawson engraving (Fig. 8), but the plan shows bracing that was not derived from either the Lawson or the Busby views. The text that accompanies the plates suggests a firsthand investigation of this bridge and especially of its details (see Fig. 11).

It should be noted that, while the arched ribs appear to have been working as compression members, the trussing posts closest to the abutments would have introduced a significantly high vertical shear load on the ribs.

Illustration courtesy of the Smithsonian Institution, National Museum of American History.

14

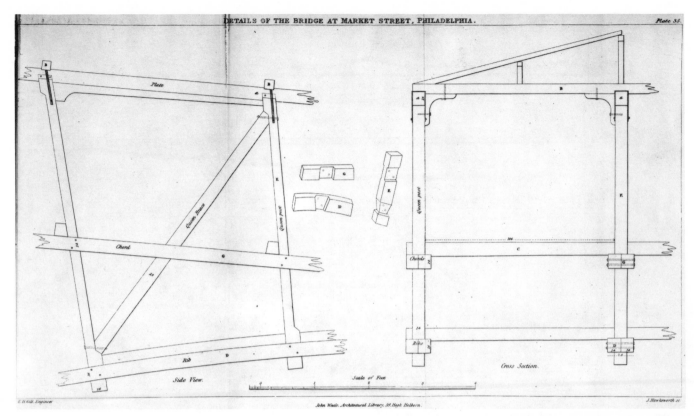

FIG. 11.

"Details of the Bridge at Market Street, Philadelphia," drawn by E.H. Gill, Engineer, engraved by J. Hawksworth, and published as Plate 35 in John Weale's *Bridges in Theory, Practice and Architecture* (London, 1839–1847). This view of the "Permanent" Bridge is interesting because it shows a number of details that are not shown in the Lawson engraving and some that are at odds with the Busby drawing. This view shows the bottom member as a multiple-membered arched rib comprised of paired 14 in. by 16 in. timbers secured with two bolts to form a composite member measuring 16 in. by 28 in. and relieved to receive what were called "radiating queen posts."

Illustration courtesy of the Smithsonian Institution, National Museum of American History.

15

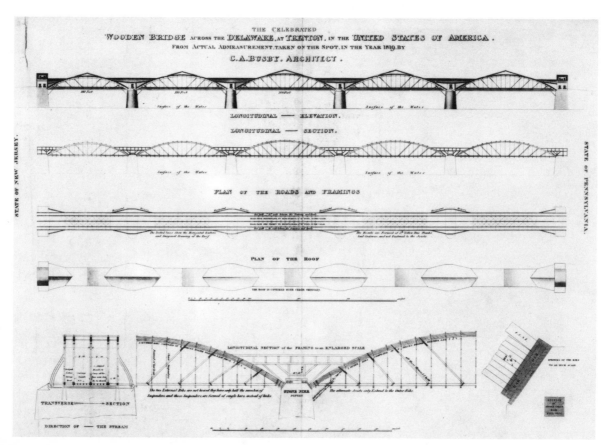

FIG. 12.

"The Celebrated Wooden Bridge Across The Delaware. At Trenton. In The United States Of America. From Actual Admeasurement. Taken On The Spot. In The Year 1819. By C.A. Busby. Architect." This bridge was built in the years 1804–1806 by Theodore Burr. With this one bridge, Burr brought America into the mainstream of laminated arch construction. He totally eliminated all of the mortise and tenon joints that had been the technological basis for most of the prior American experience. He also eliminated the shouldered posts that were so wasteful and inefficient, and which characterized the bridges of Timothy Palmer and his contemporaries working at the turn of the century.

Burr's bridge was a true laminated arch (perhaps the first in America) with a suspended roadway. It consisted of five arches, though there is some disagreement about the exact length of the spans. Roughly, there were three at 200 ft, and the other spans were 180 ft and 160 ft. Busby cites one at 200, one at 180, and one at 160, with two not called out on the engraving. Theodore Cooper cites the spans as two of 203 ft, one of 198 ft, one of 186 ft, and one of 161 ft "in the clear." This bridge also set a new American standard for the use of iron, principally for the elongated iron suspension links that supported the roadway. These links were similar to those used by James Finley in his later suspension bridges.

The arches seen here consisted of eight white pine planks 4 in. by 12 in., *bent* to the curve, and the roadway was suspended from the arch with wrought iron "chains" (actually bars) that were 1⅛ in. square in section, and the laminations were held in place by an iron wedge or "key" passing through the link on the top of the arch. The latter detail is shown in Herman Haupt, *General Theory of Bridge Construction. . .* (New York, 1883) (first edition 1851), Plate 9. Additional (and conflicting) details are shown in an article by Theodore Cooper (see Fig. 13).

Burr was a prolific bridge builder, but most of his bridges were through trusses reinforced with arches, and this type became commonly known as "Burr Arches;" but the Trenton Bridge was an anomaly, and apparently was not repeated in his later work. Burr patented a bridge design on February 14, 1806, but it cannot be proven whether it was for the laminated arch type or the arch reinforced truss type.

Through Busby's drawing, Burr's Trenton Bridge became known to European and American engineers, but it seems to have had little influence upon American bridge building, and this might suggest that this was due to Burr's patent. If so, perhaps the patent was for the Trenton anomaly rather than the Burr arch reinforced truss. This bridge survived with modifications until 1886.

Illustration courtesy of the Smithsonian Institution, National Museum of American History.

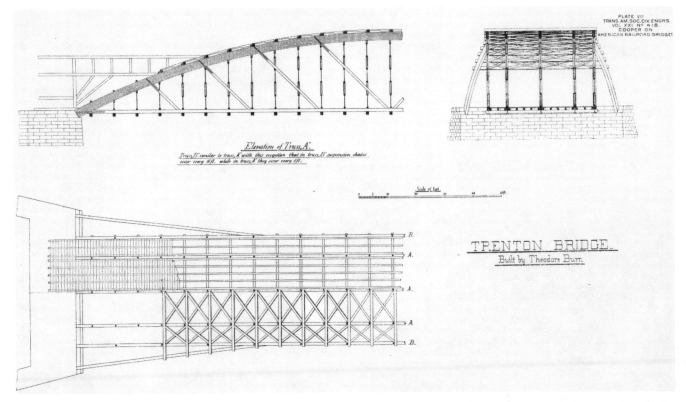

PLATE VII.
TRANS. AM. SOC. CIV. ENGRS.
VOL. XXI Nº 418.
COOPER ON
AMERICAN RAILROAD BRIDGES

Elevation of Truss, A.

Truss B similar to truss, A with this exception that in truss B suspension chains occur every 16 ft. while in truss A they occur every 8 ft.

Scale of feet.

TRENTON BRIDGE.
Built by Theodore Burr.

FIG. 13.
This view of the Trenton Bridge, by Theodore Cooper, though made much later than the Busby drawing, shows several important details rather clearly, including the laminated arch and the suspended roadway, the latter aspect being comprised of suspension ''chains,'' which were in actuality wrought iron bars forged into long links with a square cross section. In this drawing, the links were shown suspended vertically through the arch, whereas the Busby drawing shows them penetrating the arch *normal* to the arch. The Cooper drawing is also different with respect to the diagonal bracings, in that they seem to intersect the roadway beams at the panel points; but the Busby drawing shows a more random relationship between the diagonal braces and the panel points. These diagonals were intended to prevent the ''wave'' action common to suspended roadways; though we might wonder where Burr obtained his experience in this matter. We are not aware of any prior American instance utilizing a suspended roadway.

Plate VII, in the article by Theodore Cooper, ''American Railroad Bridges,'' in the *Transactions* of the American Society of Civil Engineers, ASCE, no. 418, vol. XXI, July, 1889. Courtesy of the Smithsonian Institution, National Museum of American History.

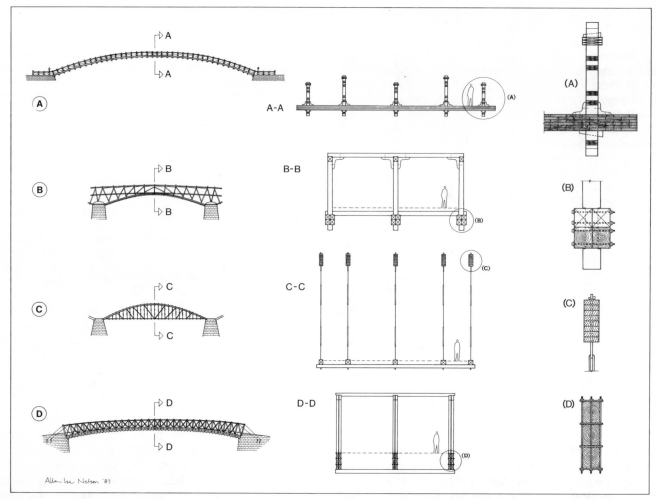

FIG. 14.

Drawing showing Comparative Aspects of four American Wooden "Laminated" Arches: Bridge Design and/or Construction between 1797 and 1812.

BRIDGE "A": Proposed design by Charles Willson Peale. *Date/Place:* 1797, proposed for the Schuylkill River, at High Street, Philadelphia. *Span:* Single span 390 ft between abutments. *"Laminated" Rib Data (Detail A):* Laminations were to consist of 2-in. thick layers bent to form a single laminated deck 1-ft thick and 50 ft wide. The laminations were to run lengthwise with the 390-ft span and were to have been held together with locust treenails. *Total Cross-Section at Abutment:* 7200 in².

BRIDGE "B": Designed and built by Timothy Palmer. *Date/Place:* 1801–1805, across the Schuylkill River, at High Street, Philadelphia. *Span:* Three spans, the center span at 194 ft 10 in., and the two side spans at 150 ft. *"Laminated" Rib Data (Detail B):* Each span consisted of three arched (and trussed) ribs; each rib was composed of paired 14 in. by 16 in. members, creating three composite ribs, each measuring 16 in. by 28 in., which were bolted together. *Total Cross-Section at Abutment:* 1344 in².

BRIDGE "C": Designed and built by Theodore Burr. *Date/Place:* 1804–1806, across the Delaware River at Trenton, New Jersey. *Span:* Five spans, three of which were about 200 ft, one span at about 160 ft, and one at about 180 ft (there is no consistent or reliable information about the exact lengths of the spans). *"Laminated" Rib Data (Detail C):* Each span consisted of five arched ribs, each composed of eight white pine planks 4 in. by 12 in. bent to the curve of the ribs and held together by iron wedges through the top loop of the iron suspension links that penetrated the arched ribs at each panel point. *Total Cross-Section at Abutment:* 1920 in².

BRIDGE "D": Designed and built by Lewis Wernwag. *Date/Place:* 1812–1813, at the Upper Ferry crossing of the Schuylkill River, Philadelphia. *Span:* Single span at 340 ft. *"Laminated" Rib Data (Detail D):* This single span consisted of three arched (and trussed) ribs, each composed of six members varying in size from 6 in. by 12 in. at the center to 6 in. by 16 in. at the abutments. The individual members were cut to the curve of the ribs and bolted together while being spaced apart by horizontal iron links and vertical iron bolts. *Total Cross-Section at Abutment:* 1728 in².

Comparative drawings by Allan L. Nelson, 1987.

18

The "Colossus": An Historical Outline

Wernwag's own approach to long-span arched bridges and the daring application of his design for the 340-ft clear span "Colossus" across the Schuylkill River near the Waterworks in Philadelphia in 1812, coupled with the high level of its historical documentation, gives us a rare opportunity to evaluate the technology of a truly remarkable American bridge in its engineering context.

While its predecessor in Philadelphia, the pioneering and long-lived "Permanent" Bridge, is known to us from a justifiably proud completion report, the "Colossus" is even more replete with historical and technical information that facilitates an attempt to provide some insights into its structural and technological planning and execution.

The bridge company business records survive, not complete in every detail, but remarkable nevertheless. Furthermore, there is a company-issued completion report that includes, for example, a listing of the member sizes and weights; and there was an "official" (and now rare) aquatint and plan of the bridge (see Fig. 15). We also have Wernwag's own promotional broadside that provides invaluable structural details (see Fig. 16). In addition, there are the numerous and sometimes inaccurate other views and descriptions of the bridge in travelers' accounts, engineering treatises, encyclopaedias, etc., etc. (see Fig. 24). In short, this may be the best recorded structure of that entire transitional period of American bridge engineering.[12]

Despite numerous references to this bridge in bridge-related literature, some of its details have never been published, and there has been some misleading information about its construction that has been perpetuated by later writers. The abundant record makes it possible to clarify the various claims regarding its design, to note changes in design that occurred during construction, to discuss problems that occurred shortly after construction, and to evaluate the efficacy of its design from a structural point of view. Therefore, the following is an historical outline of the actions and events related to the capitalization and to the design and construction of the bridge, officially known as the Lancaster-Schuylkill Bridge, but variously called the Upper Ferry Bridge, Wernwag's Bridge, and, later, the Fairmount Bridge, until it burned in 1838. Actually, it probably was Wernwag who dubbed it the "Colossus." Few of his contemporaries used that term, but the name has taken hold in historical hindsight.

January 30, 1811: A proposed Act of Incorporation was printed and circulated by a small group of Philadelphia businessmen and speculators to promote public interest (and investors) in a bridge at the site of the Upper Ferry. The proposed Act called for an initial capitalization of $40,000 based upon 800 shares at $50 each.[13]

March 28, 1811: Governor Simon Snyder signed the authorizing act after the bill was passed by the Pennsylvania House and Senate. The legislative process was well established due to the many works of "internal improvements" (turnpikes, canals, and bridges) being constructed during the first decade of the nineteenth century.[14]

June 10, 1811: Governor Snyder signed the Act of Incorporation after the 800 shares were subscribed. The company then elected officers and began negotiating for property on each side of the river at the site of the Upper Ferry owned by Abraham Sheridan, who, of course, became one of the Managers of the company.[15]

September 26, 1811: The Managers issued a public notice calling for the submission of proposals by the 24th of October, and "not to have more than one pier (a single arch would be preferred)." The roadway had to be 36 ft wide in the clear. The proposals were to include a description or plan and complete costs, including the roof and abutments.[16]

October 24, 1811: The only proposal received by the deadline was from Thomas Pope, the indefatigable architect and landscape gardener/entrepreneur, who had devised what he termed a "Flying Pendant Lever Bridge," which was actually a series of interlocked and notched timbers cantilevered from a massive masonry pier (Fig. 17). He had built and exhibited a scale model of such a bridge in New York City. For this occasion, Pope also submitted a model and rather detailed specifications (which still survive). The span was to be 432 ft, of white oak with two ribs, and 46 ft above the river. The deck was to be laid longitudinally, with each plank "tabled" into the other so as to be a continuous structural membrane to help resist the weight of loaded wagons. The top edge of each rib was to be capped to secure it against the weather. The space in the abutments between the ribs could be used for warehouses or other income-producing purposes. The bridge was to be 36 ft wide at the abutments and 26 ft wide at the center, so that the bridge plan was a sweeping curve to provide wind bracing. Pope's estimate of costs was $50,000, not including the costs of roofing, building foundations for his abutments, or his services at $5 per day with lodging and boarding at the site.[17]

October 26, 1811: Two days after the deadline, two proposals were submitted by Robert Mills, the Philadelphia architect. The first proposal included a plan and elevations for a bridge *without* a roof, the timbers above the floor to be "lined or cased within as well as without, & secured from injury of rains," on the theory that the interest on the money that would have been spent on a roof might be sufficient to renew the floors when needed. Mills' second proposal, with a plan and elevations, was for a bridge with *covered* passageways. The base price for the first proposal (without a roof or toll house) was $38,564. The designs must have had differences other than the covering, because he estimated that a roof on the first would add $7,500, bringing the total to $42,574. The designs have not survived, but we conclude that they were different. Mills' letter does not reveal whether his designs were for a single span, or a double span, or both. He later exhibited his designs at

cont'd on page 24

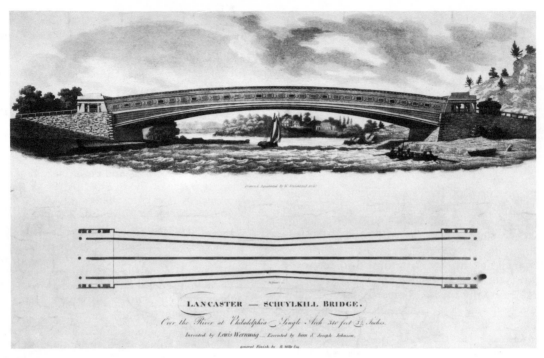

LANCASTER — SCHUYLKILL BRIDGE.

Over the River at Philadelphia — Single Arch 310 feet 3½ Inches.
Invented by Lewis Wernwag — Executed by him & Joseph Johnson.
general Finish by R. Mills Esq.

FIG. 15.
"Official" Plan and Elevation of "Colossus" (i.e. Lancaster Schuylkill Bridge), drawn and aquatinted by William Strickland, Architect, and engraved by William Kneass, Philadelphia, 1814. This view was officially commissioned by the Managers of the bridge company. On March 24, 1814, their *Minutes* reflect that, "a handsome drawing of LS Bridge executed by Wm. Strickland, by order of the board, was laid before the managers, and approv'd of. And on motion it was resolv'd, That the same be engraved by Wm. Kneass, in the best manner. . ."

Several weeks later the secretary directed Kneass to print 180 copies and an additional twenty others "handsomely colour'd." It is interesting to note that it took the Managers several drafts before they could agree on the language for their "official" engraving, language that would precisely delineate credit for the design and construction of their bridge. The first draft read, "Plan by Lewis Wernwag, Executed by Lewis Wernwag and Joseph Johnson, and the general finish of the covering by Robert Mills." Two more drafts were required before the Managers finally agreed on the simpler version which accompanies this engraving, "Invented by Lewis Wernwag—Executed by him & Joseph Johnson. general Finish by R. Mills Esq." Considering all the confusion over the years about who actually designed this bridge, especially the frequent misattribution to Robert Mills, this inscription sorts it out succinctly and precisely!

It is further unfortunate that this print is so rare and little known, because it is the only published source that correctly shows the plan of the three ribs, with the outer ribs flared to provide built-in wind bracing (the better known Busby drawing of 1823 shows the ribs parallel, which was incorrect). The ribs were 13 ft 1 in. apart at the center and 21 ft apart at the abutments, the transition (in plan) being straight-line segments rather than a continous curve. The two outer footways were 3 ft wide at the middle and 4 ft wide at the abutments.

This view shows the housing, the entry porticos, the roof, and the circular toll booth, all designed and built under contract with Robert Mills, an architect then practicing in Philadelphia but sometimes working as a construction contractor to supplement his income. It was Mills' design of the covering that helped make this daring structure into a soaring example of classical revival design as applied to bridge construction.

Mills' design differentiated the curve of the supporting ribs from that of the roadway. A series of window openings lighted the interior. They were interspersed with solid framed panels to continue the rhythm of the window openings. The splayed piers at the porticos, with arched openings and classical columns, were architecturally less convincing than the way Mills articulated the structural curves of the ribs and the roadway.

Some of these details may have been removed as a result of work carried out after a hurricane in 1821, during which the roof and weatherboarding were torn off the bridge. Later views seem to suggest a somewhat simpler aspect. It is unfortunate that, because of its premature destruction by fire in 1838, the bridge did not survive into the age of photography, so that we might have a more accurate perception of its structure, its daring, and its grace—described by the English actress, Fanny Kemble (in 1832) as ". . .particular light and graceful in its appearance; at a little distance, it looks like a scarf, rounded by the wind, flung over the river."

Illustration from the writer's collection.

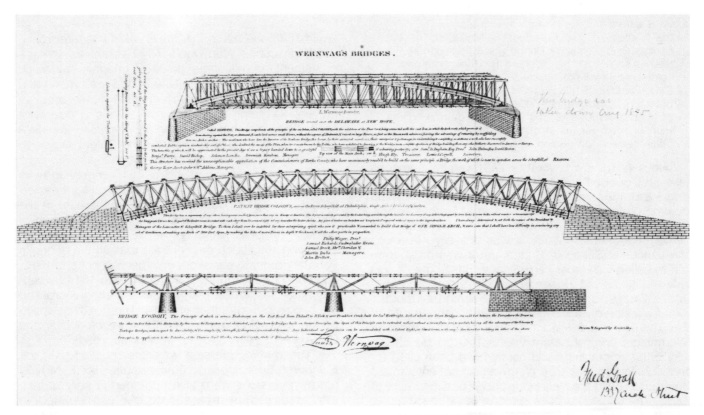

FIG. 16.

"WERNWAG'S BRIDGES," drawn and engraved by Enoch G. Gridley, ca. 1815. There is little doubt that this engraving was commissioned by Wernwag for use as a promotional broadside. The engraving includes a reproduction of his distinctive signature, together with phraseology written in the first person singular. The broadside further directs any interested individuals to the "patentee" at the Phoenix Nail Works in Chester County, Pennsylvania, of which he was manager from 1812 to about 1818.

The engraving includes three of Wernwag's bridges. The lower drawing depicts the "BRIDGE ECONOMY," a light timber drawspan which appears to have been Wernwag's first venture into bridge design (c. 1810), and, by his own claim, this "principle" was used for 52- and 60-ft spans across Neshaminy Creek and Frankford Creek, both northeast of Philadelphia.

The upper drawing shows an elevation of the "NEWHOPE," a trussed bow-string arch consisting of laminated wooden arches and numerous iron members. This bridge, constructed in 1813–1814, consisted of six equal arches of 175 ft each for a total length of 1050 ft between abutments, across the Delaware River at New Hope, Pennsylvania.

The middle drawing is Wernwag's "PATENT BRIDGE COLOSSUS." We are fortunate that such a high quality engraving exists for this bridge because it is replete with considerable information about its configuration, including the relationship between the primary supporting arch, the arch of the roadway, and

the arch of the upper chord, each of which is different from the other. In addition, the engraving provides important evidence of construction joinery, disposition of the iron tie rods, iron "links," iron eye bolts, and iron "hanger braces," some details of which were separately delineated above and to the left of the bridge drawing (also see Fig. 21). Such details must have been important to Wernwag, and he surely supplied this information to the engraver.

Given the known information about the structure (from the 1814 Report of the Managers), it is possible to carefully analyze the geometry of the engraving, which reveals a surprising accuracy as to the length of the main span in relation to its rise, as well as the number of bays, and the lengths of the king posts. The analysis also reveals, however, that the individual members have been rendered oversized, probably to allow the engraver to show such details as the joint connections and the ironwork.

Wernwag's accompanying description reads, in part, as follows:

"This Bridge has a superiority of any other, having near 100 feet span, more than any in Europe or America. The dry rot is entirely prevented by the timber being sawed through the heart, for the discovery of any defect & kept apart by iron links & screw bolts, without mortice or tennon, except the king posts & truss ties. No part of the timber comes in contact with each other, & can be screwed tight at any time when the timber shrinks. Any piece of the timber can be taken out & replaced if required without injury to the Superstructure. I have always determined to set forth

cont'd on page 22

Fig. 16 cont'd

the names of the President & Managers of the Lancaster & Schuylkill Bridge [its official name]. To them I shall ever be indebted for their enterprising spirit, who saw it practicable & consented to Build that Bridge of ONE SINGLE ARCH, & sure I am that I shall have less difficulty in convincing any set of Gentlemen, of making an Arch of 500 feet span, by making the Ribs of more Pieces in depth & thickness, & all the other parts in proportion."

It is not known whether Wernwag or someone else first dubbed it the "Colossus." Because of this engraving, the term "Colossus" has become the most common reference to Wernwag's most famous bridge. It must be admitted, however, that this name was seldom used in its heyday. It was officially known as the Lancaster-Schuylkill Bridge, but it was most commonly called the Upper Ferry Bridge. Sometimes it was referred to as the Upper Schuylkill Bridge, or Wernwag's Bridge, and finally as the Fairmount Bridge.

A version of this engraving also became a part of the reconstructed Patent Office records. Although Wernwag's original patent was issued on March 28, 1812, this engraving could hardly have been used for that submission, because Wernwag did not enter into a contract to build the New Hope bridge until February of 1813. The engraving does seem to have become part of the later record of "restored" patents in connection with Wernwag's second patent of December 22, 1829. Although the Patent Office records were destroyed by fire in 1836, the *Journal of the Franklin Institute,* vol. v, 1830, describes Wernwag's 1829 patent as an "improvement" on his earlier patented bridges, "by which the floor, sides, and top of the bridge are braced and held together by diagonal, horizontal iron braces, to prevent its being crooked. . . ." Thus it does not seem likely that this engraving accompanied the 1829 patent, despite its later use by the Patent Office to reconstruct their records after the fire.

An earlier version of this c. 1815 engraving (with less descriptive text) was published two years earlier, probably in 1813. This assumption for the date of the earlier version is based on a portion of the text which reads: "BRIDGE, erecting at NEW HOPE. 1813." Wernwag concluded the contract to build the bridge in February of that year, work commenced in April, 1813, and continued throughout 1813 and into 1814 (except for winter suspension); and the first carriages crossed the bridge in September of 1814. The c. 1815 version of the engraving reads: "BRIDGE *erected* over the DELAWARE NEW HOPE" [italics supplied].

The c. 1813 version was later published in the *Edinburgh Encyclopaedia* (Philadelphia, 1832). Partially redrawn versions were also published in several mid-nineteenth century treatises on civil engineering.

The copy of the c. 1815 engraving seen here is reproduced courtesy of the Free Library of Philadelphia.

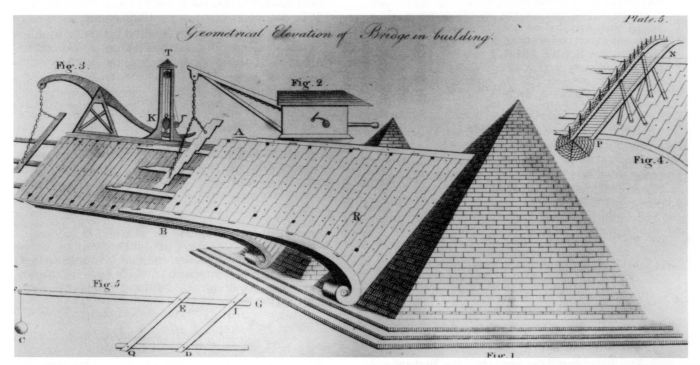

Geometrical Elevation of Bridge in building.

Plate 5.

Fig. 3. Fig. 2. Fig. 4. Fig. 5. Fig. 1.

FIG. 17.

In 1807, Thomas Pope, who described himself as an architect and landscape gardener, then of New York, patented his fantastic "Flying Pendent Lever Bridge" and then proceeded to promote it by advertisement and later by an 1811 book with an impressive list of subscribers and complete with an eight-page poem in rhyming couplets. Actually, Pope's 288-page book is the most comprehensive early American publication on the subject of bridges. It includes a 190-page section on the "History of Bridges," a 30-page section on the strength of timber, and a 50-page section on his own fantastic bridge, complete with testimony from New York shipwrights, based upon the examination of his "Grand Model." His half model was nearly 50 ft long and was reputed by Pope to have supported ten tons, which "astonished the mind of every beholder."

Pope seems to have come to Philadelphia in the latter part of 1811, and he exhibited his model at his school on Library Street, for which he charged an admission of fifty cents. He also submitted proposals for several bridges, including one at Gray's Ferry over the Schuylkill and one over the Susquehanna. He also exhibited designs for additions to the old State House (Independence Hall). None of these seems to have been successful.

Pope's was the only proposal for the Lancaster-Schuylkill (or Upper Ferry) Bridge to have been received by the Managers before the deadline of October 24, 1811. Though the drawings have not survived, it was undoubtedly a version of his Flying Pendent Lever Bridge, with a span of 432-ft, of white oak, with two ribs, and 46 ft above the water, 36 ft wide at the abutments, and 26 ft wide at the center. His estimate of costs was $50,000, not including the costs of covering nor of the foundations or his cost of superintendence.

This plate from his 1811 book is presumed to have been similar to his proposal for the Lancaster-Schuylkill Bridge, which does not seem to have been seriously considered by the Managers of the bridge company. After the contract was given to Lewis Wernwag, Pope claimed patent infringement, by virtue of Wernwag's concave sides of his bridge to supply built-in wind bracing. Wernwag avoided this by providing straight-line segments for the plan of his bridge (see Fig. 15).

This illustration is Plate 5, in Thomas Pope, *A Treatise on Bridge Architecture. . .* (New York, 1811). Courtesy of Eric DeLony.

23

cont'd from page 19

the Pennsylvania Academy of Fine Arts, noting that they had a "Span [of] 330," but such a description was frequently used for the total length even when there were multiple spans. It is likely that Mills' plan called for two arches because of a resolution dated November 14 (see below).[18]

October 26, 1811: On the same day that Mills presented his proposals, the Board of Manager's *Minutes* noted that several proposals accompanied with plans and estimates were presented and read and ordered to lie upon the table.

November 14, 1811: The bridge Managers resolved that the bridge "shall have three hundred & fifty feet Water way and two equal arches, with one pier." On this same date, however, Lewis Wernwag submitted a proposal to build a bridge with the necessary abutments and piers "according to his plan," with an incentive to keep the cost below $40,000, whereby if built for less than that, the company would split the difference with him; and should it cost more, he would give half of his wages ($5,000) for superintendence to the company.[19]

December 5, 1811: On this date, the company entered into Articles of Agreement with Wernwag, which had obviously been the subject of negotiation: "Whereas the said Lewis Wernwag hath furnished. . .a Plan of a Permanent Bridge. . .and hath proposed to superintend. . .to collect all the materials necessary. . .to find fit and suitable Workmen to do all the work. . .and to lend all his Tools and apparatus for building erecting raising and finishing of the said Bridge. . . . In consideration of all which services to be rendered. . .and for the Plan of the said Bridge and the use of his Tools and Apparatus,. . . the Company will pay within one year from cornerstone laying, the sum of $3000, with a "compliment" of $500, if they are satisfied with the superintendence, care, management and conduct of Wernwag." Interestingly, Robert Mills was a witness to the signing of this agreement, but there was no mention of the span or configuration of the bridge.[20]

December 20, 1811: The company rescinded the November 14 resolution relating to a bridge with two arches and resolved that "Mr. Wernwags plan of a Bridge with one arch of 330 feet chord be adopted." The stockholders approved the single-arch plan on January 6, 1812.[21]

April 28, 1812: There was a cornerstone laying in the eastern abutment, celebrated with full Masonic ceremonies and a hogshead of whiskey for the workmen. Many years later (ca. 1875) a copper cornerstone plate was found, and the full inscription was published in Scharf and Westcott's *History of Philadelphia.* The plate listed the names of the Board of Managers ". . .and Louis [sic] Wernwag architect."[22]

April 30, 1812: Wernwag was directed to have all of the iron castings made at the Eagle Furnace, on the banks of the Schuylkill near Callowhill Street, belonging in part to Samuel Richards (also a manager of the bridge company). Wernwag's estimate of iron (both cast and wrought) for the bridge totaled over 46 tons, an uprecedented use of iron at the time for an American bridge![23]

May 14, 1812: Wernwag was directed (by the Managers) "to have all the planed timber for the Bridge work, varnished over to preserve it provided the expense not

exceed $300." It is likely that no decision had yet been reached regarding the roofing and siding.[24]

June 12, 1812: On this day, Thomas Pope advertised in the Philadelphia *Aurora* that Lewis Wernwag "hath of late erected a model to illustrate to the public the kind of Bridge intended to be built over the river Schuylkill. . .which model doth partake of part of the. . .valuable properties belonging to my invention, namely, that the external perpendicular sides of said Model are built in the shape of two concave circles or arcs; their convex sides, of course, face each other, by which important shape, a more perfect resistence [sic] is furnished against the pressure that wind and tempest afford on the sides of a bridge. . .the advantage of which being wholly secured to me; Therefore, the adoption of this shape. . .is a direct infringement on my patent right." Pope required that Wernwag "alter, amend, or wholly destroy the aforesaid Bridge Model, so that it shall cease to partake of, or infringe on my patent right." In the long tradition of patent evasion, Wernwag did modify the plan of his bridge so that the center of the bridge was narrower than at the abutments, but instead of sweeping curves in the transition of the plan from the abutments to the center of the bridge, he narrowed it with *straight line* segments. A sketch to that effect is in the *Minutes* of the Managers (see further discussion below).[25]

June 23, 1812: Wernwag was authorized to "build his machine for driving the piling as soon as possible."[26]

July 2, 1812: Wernwag "represented to the Board that three ribs instead of five, now exhibited in the model will be sufficient for all the purposes of permanence & stability of a good Bridge." The managers agreed to three ribs, providing that Wernwag should add the extra ribs if necessary.[27]

September 24, 1812: The managers resolved that Wernwag devote "his whole attention to the raising of the Scaffolding, on which the ribs of the Bridge are to be framed." It should be noted that, a few weeks prior to that resolution, Wernwag had purchased a one-quarter interest in the Phoenix Nail Works at French Creek and was beginning to take over as Manager of that industrial operation.[28]

October–December, 1812: The company was concerned about a shortage of funds and the need to increase the capital stock to finish the bridge and build a toll house. Wernwag applied for permission to "remove his family and himself" to the nail works at French Creek. Permission was not granted.[29]

December 31, 1812: The managers noted that Wernwag was in violation of his contract, having "absented himself from the Service of the company," and that there should be a deduction in his wages.[30] However, Wernwag obviously returned for the big event of "striking" the centers.

January 7, 1813: On this day, "in the presence of the Board of Managers and an immense concours [sic] of Citizens the centres of the Bridge were struck." Many years later, Wernwag's son, John, related the event as one where "thousands" of spectators had assembled on the riverbanks expecting to see the bridge collapse upon removal of the blocking from between the bridge and the scaffolding. The managers must have had their own private misgivings considering the daring of the undertaking. Unbeknownst to them, Wernwag had secretly removed the blocking between the scaffolding and the arched ribs the day before. When the inspection team of managers approached the first set of blocks, the managers said, "Well, Lewis, do you think our bridge will stand the test today?" His reply was "Yes, gentlemen, it will." When they discovered that all of the blocks had been removed and that the entire bridge was free and clear of the falsework, Wernwag related to his son that he never saw the countenances of men brighten up as those of the managers when he informed them that the bridge was freed from the scaffold the day before. As of this date, some $64,500 had been spent, including the costs of property and to "extinguish the right of ferriage" and numerous other miscellaneous costs. The remaining costs were estimated to be about $20,550.[31]

March 13, 1813: The managers resolved "that it is expedient to roof and finish the Bridge. . ." Tolls were being collected in a temporary toll house. By March 26, the committee appointed to contract for covering the bridge had entered into a verbal agreement with Robert Mills for covering and enclosing the bridge, as well as building a toll house, all to be completed in six months for a cost of $4520, not including materials. The roof was to be shingled with 3-ft shingles, 10 in. to the weather, the flanks to be studded and boarded with quartered boards, planed both sides, tongued and grooved. There were to be ten window openings with "blinds to slide" on each side of the bridge. The "abutment buildings" were to "embrace the extent of the Iron Bars secured to the abutments, opened & finished with Columns." The toll house was to be built with wood, the roof shingled, and a "Colonnade surrounding the walls."[32]

Early May, 1813: At about this time, the bridge (actually, its western abutment) and northern wing wall began to exhibit some movement that for a time threatened the stability (and safety) of the entire undertaking. In the following weeks, a number of "experts" were called in for consultation, including Robert Mills and Oliver Evans. For Evans' report, see "The Abutments and the Structural 'Defect'" this book. Robert Mills' roofing contract was put on "hold" during the several months of this structural crisis.[33]

July 31, 1813: After the Managers had agonized over the alarming structural movement and called in experts for consultation, and after they called Wernwag back to view the problem and share their distress, they decided to lay the entire matter before the stockholders on July 31. For that report, see "The Abutments and the Structural 'Defect'" this book. Subsequently, some experimental remedial work was carried out; and, finally, after the situation stabilized, Wernwag cut away the stone behind the bearing of an arch, bringing it back into alignment. Mills' roofing contract was once again ordered to be resumed.[34]

September 23, 1813: The managers resolved that $200 be presented to Joseph Johnson "as a mark of gratitude and compensation. . .for his diligence and intelligence in remedying an unfortunate defect in the Superstructure of the. . .Bridge." Johnson appears to have been Wern-

wag's principal assistant; and they were to collaborate on several subsequent contracts, including bridges at New Hope and Pittsburgh.[35]

November 4, 1813: Robert Mills reported that the "sanding" in the paint on the Market Street Bridge "seemed to be decaying and that he recommended a fourth coat of paint without the expense of sand," which would save $400.[36]

December 23, 1813: The toll house roof was being covered, not with shingles, but with sheetiron (painted both sides), and Mills' contract was substantially completed.[37]

January 6, 1814: The managers were happy to assure the Stockholders that "to the best of their belief the Bridge is perfectly solid in all its parts. The defect in the upper [upriver] and middle ribs, which gave some uneasiness [read understatement] last Summer, is now remedied, and the inward pressure of the western abutment stopt."[38]

March 3, 1814: The managers published 500 copies of a rather detailed *Report of the Managers of the Lancaster and Schuylkill Bridge Company to the Stockholders.* A major extract of this report was included as an appendix to this writer's article "The Colossus of Philadelphia," published in *Material Culture of the Wooden Age,* edited by Brooke Hindle, Sleepy Hollow Press, Tarrytown, NY, 1981, pp. 159–183.

March 24, 1814: A "handsome drawing" executed by William Strickland was "laid before the managers," who resolved that it be engraved by William Kneass. Subsequently, 180 copies were executed, and, in addition, 20 others were to be "handsomely colour'd," one of which was to be framed and presented to the Governor of Pennsylvania. However, on April 30, the managers were still discussing the wording to accompany the engraving. The wording for the first draft read as follows: "Plan by Lewis Wernwag, Executed by Lewis Wernwag and Joseph Johnson, and the general finish of the covering by Robert Mills." In the *Minutes* of the managers, that version was crossed out and replaced with the following: "The Design in general new, invented by Lewis Wernwag; executed by him and Joseph Johnson: general finish by Robert Mills." This is the version (omitting "The Design in general new") that was published with the aquatint (see Fig. 12).[39]

June 8, 1814: The managers determined that Mills' carpentry work on "frame Roofing and weatherboarding has not been done according to contract and has been badly executed." Eventually, the matter was handled by third party "surveyors" in an attempt to arbitrate the dispute and settle the accounts. This dragged on through 1815, and it does not appear to have been settled; we conclude that Mills was left uncompensated for $936.[40]

September 3, 1821: The bridge suffered extensive damage when the roof and weatherboarding were "torn off" by a hurricane. William Strickland was invited to examine the state of the bridge, to prepare a cost estimate for repairs, together with the "proper form" for a newspaper advertisement to invite proposals to furnish materials to repair and reroof the bridge. Strickland produced a plan and estimate that were approved. Apparently the company did not have sufficient financial reserves for the work and had to borrow monies, and the managers had to pledge the net tolls until the advances were repaid with interest.[41]

1822: In this year, the bridge company was $43,826 in debt, including the expenses for reroofing the bridge. The company had never paid a dividend to the stockholders. The average tolls for the years 1819–1821 had been $3,124, compared to the "Permanent" Bridge at Market Street that averaged $14,813 per year, thus enabling a good level of maintenance to the roadway and the roof.[42]

November, 1828: Mr. Frederick Graff (an engineer from Philadelphia) was requested to "examine into the present state of the Bridge," and he indicated a willingness to undertake the examination if it could be deferred until the ensuing spring. The purpose of this intended examination is not known, but it did not take place. The company then sought to obtain Joseph Johnson to undertake some alterations. Finally, in June of 1830, a committee was formed to make some unspecified repairs, but perhaps that too was deferred. Those repairs, if carried out, are not specified in the *Minutes.*[43]

May 9, 1831: Lewis Wernwag attended a special meeting of the Stockholders and gave a verbal report on the state of the bridge. A committee was appointed to attend to the repairs as soon as Wernwag was prepared to commence operations. He sent a letter in mid-August giving reasons for not complying with his promise to commence the repairs. He had been living for some years on Virginius Island at Harpers Ferry (then in Virginia). The nature of the proposed repairs or their outcome is not known.

1832: The bridge was scaffolded and painted.[44]

September 1, 1838: The bridge was destroyed by fire at about 9:00 p.m., apparently by incendiaries. "The fire commenced at the west end of the bridge, and burnt with such fury, that in 20 minutes after its discovery the part laying on the abutment was burned off, when the remaining part fell into the river. The Toll house was also destroyed."[45]

1842: Wernwag's "Colossus" was replaced with an equally impressive single-span suspension bridge, a pioneering structure of woven wire rope designed by Charles Ellett, Jr.[46]

The "Colossus" Superstructure

We know (from references in the *Minutes* and the bridge records, cited previously) that significant design changes were made by Wernwag just prior to construction. For example, he reduced the number of ribs from five to three, and he subtly changed the shape of the plan of the bridge (basically, wider at the abutments than at the middle) from sweeping curves to straight line segments "pinched" at the middle. We also know that some changes were made to the trussing details shortly after completion of the bridge when alarming movements occurred in the western abutment, causing the two upriver ribs to be raised, thus throwing the deck out of level. We do not know the exact nature of those changes, but the overall aspects of the construction are well documented, and it is the design and construction of the superstructure that are discussed in this section.

As completed, the basic superstructure of Lewis Wernwag's "Colossus" consisted of three multiple-membered (or laminated) arched ribs trussed with numerous king posts, diagonal braces, and counterbraces and longitudinal truss ties all designed as a *system* with paired members; that is, the ribs were composed of paired members, as were the king posts and the truss ties, and of course the bracing was "paired" to form an "X." The "pairing" seems to have been derived from Wernwag's concept that all of the major members were to be sawn through the heart (to detect unsound material and allow it to dry more quickly) and then to be matched back together but spaced apart to allow air circulation and prevent dry rot. The ingenuity of this system becomes more apparent as one studies it in the third dimension (see Figs. 18–20).

Certainly the most interesting aspects of the design were the details of the "laminated" ribs. It should be clarified at the outset that the laminations were bolted, not *through* the members but *between* the members, into what was intended to be a structural entity. Nor were the laminations built up of relatively thin boards or planks, as had been proposed by Charles Willson Peale (Fig. 6). Instead, each large rib consisted of six timbers measuring an average of 6 in. wide by 14 in. deep, and the six timbers were arranged two timbers abreast and three timbers deep, thus forming composite ribs that measured (on the average) 1 ft wide and 3.5 ft deep. Furthermore, none of the "laminations" were bolted through the wood to hold them in place. Instead, they were spaced apart (about 1 in.) horizontally with iron spacers called "links" and vertically with heavy bolts that spaced the timbers vertically and passed through the links (Fig. 19).

It is not known whether the timbers were cut to the curve or bent to the curve, but there are several reasons why we can assume that they were cut to the curve. There is mention in the *Minutes* to the timbers having been "planed," and we believe that that was the process used to dress the timbers to the curve. Furthermore, while it would have been technically possible to bend such large timbers on their vertical axis, it would have required considerable force, and it would have been difficult to keep them all in parallel alignment. That would have amounted to prestressing the timbers and would have made them extremely difficult to remove, and removeability for replacement was one of the advantages claimed by Wernwag. If bent to the curve, any replacement member would have to be prestressed to be able to fit it into the space of the member which was to be removed; furthermore, the defective member would have to be prestressed in order to relieve it of its load prior to removal. Finally, even more convincing is the fact that similar "laminated" ribs built into the New Hope Bridge (being built concurrently by Wernwag and Johnson, though started a year later) were *dressed to the curve of the arch* (italics supplied). It should be noted that the New Hope ribs were approximately the same size, measuring 6 in. by 15 in. whereas the "Colossus" ribs varied from 6 in. by 12 in. to 6 in. by 16 in. Fortunately, there is a good published completion report for the New Hope Bridge that specifically mentions the "dressed" ribs.[47]

One of the claimed advantages of the ribs with their laminations spaced apart was that no timbers were in direct contact with each other except at their ends. Wernwag claimed:

> No part of the timber comes in contact with each other. It can be screwed tight at any time when the timber shrinks. Any piece of timber can be taken out & replaced if required without injury to the Superstructure.[48]

How he would have removed any of the ribs after completion of the covering and the flooring is not clear. In fact, it seems highly problematic, considering the geometry of the *completed* bridge. However, we know that Wernwag's claim regarding the "ease" of removing any damaged or rotted rib member was "experimentally" (and successfully) carried out *during* the construction of his bridge at New Hope, which was well underway before the "Colossus" was finished.[49] It would seem to this writer that removing such a timber *after* completion would have been a virtual impossibility.

Another structural subtlety was that Wernwag varied the depth of the ribs while maintaining a constant width. At the abutments, the laminations were 16 in. deep, while at the center they were only 12 in. deep. Thus the total cross-section area of each rib was 4 ft^2, and at the center it was 3 ft^2.

Each laminated rib was set in large cast iron "head blocks" at the abutments. These would have

cont'd on page 31

27

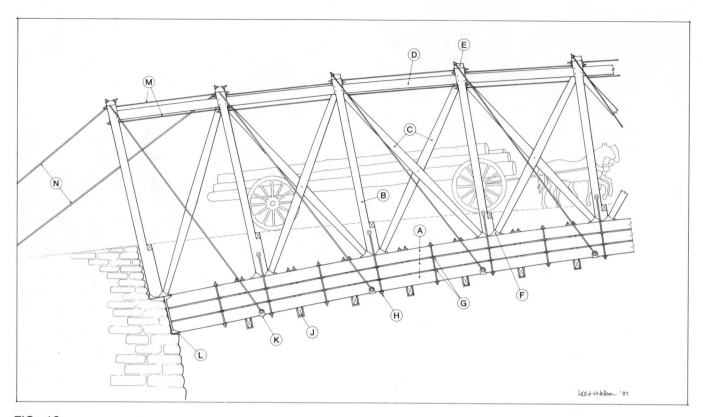

FIG. 18.
Scale drawing of the superstructure, consisting of a partial elevation, based upon many known facts but also presented with certain assumptions identified below. It should be noted that there is a great deal of information in the bridge company records, in the 1814 Report, and in the Gridley engraving (Fig. 16). However, there are some inconsistencies, and some of the information requires a certain amount of interpretation. For example, the king posts were described as being 1 ft square in the Report for purposes of weight calculations, but in reality they consisted of two 6 in. × 12 in., made from a 12 in. × 12 in. timber sawn down the middle ("slit through the heart, so as to show any defect"). What follows then is a detailed description of the major features of the superstructure related to the drawing with an alphabetical "key," using the nomenclature in the 1814 Report, where possible:

"Structural Feature "A": consists of 3 Laminated Ribs, each consisting of 6 white pine timbers (2 timbers wide and 3 timbers deep), each timber measuring 6 in. × 16 in. at abutments, 6 in. × 14 in. at the quarter span points, and 6 in. × 12 in. at mid-span, so that the "average" rib measured 12 in. wide and 3 ft-6 in. deep (4 ft-0 in. deep at the abutments). All timbers were spaced apart with 1 in. iron bolts and wrought iron "links," thus "being prevented from coming into contact, is secured from the dry rot." Wernwag asserted that, as these timbers dried, the nuts could be tightened to accommodate the shrinkage. While we know the sizes of the rib members, we do not know their lengths. We estimate that they averaged 35–40 ft in length, probably with staggered joints. Although Wernwag claimed that any individual rib could be replaced, and while that may have been theoretically possible, it is doubtful that such replacement would have been feasible or practical. To do so would have required removal of numerous bolts, links, stirrups, cast iron boxes, and perhaps diagonal iron bars, to say nothing of the almost impossibly restrictive constraints of the flooring, the side wall studding and side wall enclosure. Such a claim was almost certainly an exaggeration. However, at the New Hope Bridge (built by Wernwag a year later), the removal of a rib was "fully and experimentally ascertained during the raising of the arch." At that stage of construction, there were no constraints of flooring, siding, etc.

There is no specific evidence to clearly establish whether the ribs on this bridge were *bent* to the curve of the arch or *cut* to the curve, but we believe that they were cut to the curve for several reasons. First, there are references in the *Minutes* to timbers having been "planed" and the need to have them "varnished over" to preserve them. This occurred early in the construction and was likely for the rib members. Second, the dimensions of the ribs (6 in. × 16 in.) would have required considerable force to bend and to hold in

cont'd on page 29

28

Fig. 18 cont'd

place during construction, compared to bending the 4 in. × 12 in. used on Burr's Trenton Bridge. Another reason for believing that the ribs were cut to the curve is that Wernwag's New Hope Bridge is *known* to have been made that way. The accounts of that bridge say that the ribs were "dressed to the curve of the arch," each rib being 6 in. × 15 in.

Structural Feature "B": consists of 87 doubled king posts (29 doubled king posts on each of the three ribs), of white pine, each king post being identified in the 1814 Report as 1 ft square (for weight estimating purposes), but actually they were sawn down the middle, giving them a nominal size of 6 in. × 12 in., the two halves then being paired and spaced apart. It is assumed that they were 6 in. × 12 in. at the top where they aligned with the 12-in. width of the 8 in. × 12 in. caps or cross ties. The king posts were then necked down for most of their length, possibly to 6 in. × 10 in. (as drawn here). This gives a smaller number of cubic feet than if the posts were 12 in. × 12 in. for their full length, but Wernwag may have simply rounded off the volume in his tabulation, knowing it would be compensated by the floor beam shoulders at the bottom.

The king posts are listed in the 1814 Report as being 15 ft long, which was the *average* length (at the quarter-span points), ranging from about 13 ft 6 in. long at mid-span and 21 ft 6 in. long at the abutments. These lengths are based upon calculations from the engraving, and they agree with the average lengths for both the king posts and the braces, as given in the Report.

Structural Feature "C": consisting of 162 braces and counter braces (diagonals between the king posts), white pine. There are 54 braces on each of the three ribs, as seen on the Gridley engraving; but the 1814 Report calls for 84 braces, each 1 ft square, which appears to have been calculated as though there was one 12 in. × 12 in. brace in each of the 28 bays, whereas there were double braces in each of 26 bays and one brace in each of the end bays, based on the apparent arrangement seen in the Gridley engraving. It is possible that the two end bays had *paired braces* in the same alignment, which would make a total of 168 rather than 162 braces as assumed. The cross braces are similarly assumed to have been sawn down the middle to form 6 in. × 12 in. braces, spaced apart to align with the king posts. The braces are identified in the 1814 Report as being 18 ft 9 in. long, which is their *average* length at the quarter-span points. The Gridley engraving shows the braces slenderer at the ends than at the middle and pegged at the crossing of the braces. We have assumed that to be the case for this drawing.

Structural Feature "D": consisting of 84 doubled truss ties (28 doubled ties on each of the three ribs), white pine. The 1814 Report notes that there were three truss ties, 343 ft long by 1 ft square; but that was a simplification for weight estimating purposes. Actual-

ly, the doubled set of truss ties were about 12 ft long between the tops of the king posts, but they essentially comprised a timber 343 ft long.

Like the king posts, the 12 in. × 12 in. truss ties were cut down the middle, making 6 in. × 12 in. pieces which were then paired and spaced apart about 1 in.

Structural Feature "E": consisting of 29 caps, on top of the king posts, white pine. These were listed in the 1814 Report as being 36 ft long and measuring 8 in. × 12 in. This listed length is an *average* length. Since the total width of the three ribs was 28 ft 2 in. at mid-span and 44 ft at the abutments, the average width would require the caps to be 36 ft long and is thus consistent with the Report. It is this writer's assumption that the smaller members of the bridge (less than 12 in. × 12 in.) were *not* sawn down the middle, "to show any defect." Thus the caps have been drawn as 8 in. × 12 in. laid flatways on top of the king posts, which are assumed to be 12 in. wide at their tops. It is not known how the caps were connected to the king posts. The 1814 Report states that the only mortise and tenon joints were between the king posts and the longitudinal truss ties. It is assumed that they were notched and spiked.

Structural Feature "F": consisting of 87 cast-iron boxes, one for each set of paired king posts and diagonal bracing. The exact design is conjectural. The 1814 Report describes the king posts as being ". . .set in cast iron boxes of a proper shape to receive them. . . ." It is not known whether these "boxes" were in fact "open" boxes, that is, with recesses to receive the wooden king posts and diagonal bracing, or if they were solid castings like other contemporary "boxes" with beveled bearing surfaces. The writer has assumed the latter case for the purpose of these drawings for two reasons: (1) Although later bridge construction (by the 1830s) exhibited some rather sophisticated iron work as used on the B & O Rail Road bridges designed by Latrobe and published by von Ghega, the state of the art of iron castings in 1812 was considerably simpler. As drawn here, the castings would not have required the use of a cope and drag (a two-part mould), nor would they have required cores to produce cavities in the castings. As drawn, they would have been simple open pit sand castings, like that used for large iron stove parts or most engine parts. (2) Another advantage of solid "boxes" without cavities is that it would simplify the logistics of erecting the trussing members, especially the diagonal bracing. The bracing would have to be in place and temporarily supported before the king posts and truss ties could be assembled with their mortise and tenon joints. Actually, these cast-iron boxes might have been either of open or solid design as it was a transitional period of iron work, but there are no known surviving details of such "boxes" for that date. In addition to these castings, the primary bearing plates for the arches (at the abutments) were also identified

cont'd on page 30

Fig. 18 cont'd

as cast iron and were called "head blocks." The 1814 Report tallied a total weight for cast iron (as opposed to "bar" iron) of 11 tons. The boxes as drawn here would have weighed about 10 tons, leaving one ton for the six bearing plates, which appears to be reasonable. According to the *Minutes,* the castings were to have been made at the Eagle Furnace on the banks of the Schuylkill at the foot of Callowhill Street, belonging in part to Samuel Richards, who was also a Manager of the bridge company.

Structural Feature "G": consisting of 579 links, to "separate the timbers," utilizing 252 1-in. iron bars (bolts) between the rib members to prevent them "from coming into contact." These bolts varied in length from about 4 ft 6 in. to 5 ft 0 in. These links were separately delineated and identified on the Gridley engraving. They were also used in connection with the "eye bolts" (see below) at the king posts.

Structural Feature "H": consisting of 243 iron eye bolts (probably 1-in. bars about 7 ft long), which separated the doubled king posts and which connected the king posts to the ribs, the center bolt "going through the middle of the cast iron box." These bolts also extended through links identical to those used to separate the rib members at each mid-bay (see item G").

Structural Feature "J": consisting of 54 cross ties, 6 in. × 12 in. (average of 44 ft long), probably white pine, used in conjunction with 54 "strong iron stirrups; upon these ties, and between the ribs. . ." (from the 1814 Report). The stirrups must have been 1-in. bars to be consistent with the known sized 1-in. bars used with the links to separate the rib members.

Structural Feature "K": consisting of 84 iron hanging braces with stirrups and bolts, which extend from the tops of the caps (on top of the king posts) down to the ribs in a diagonal direction, with a bolt *through* the bottom rib and the stirrup. These braces were separately delineated and identified on the Gridley engraving. The hanging braces varied in length due to the changing height of the king posts, but their average length was about 26 ft, and they are assumed to have been of 1-in. iron, but they could have been larger because they were not used to separate the ribs.

Structural Feature "L": consisting of six cast iron head blocks, used where the laminated ribs bear against the masonry abutments. According to the 1814 Report, the three ribs, "extending across the river, from one abutment to the other, are set in cast iron head blocks, upon an offset in the wall, twelve feet above common high water mark." The details of these castings as shown on this drawing are vague and based upon the Gridley engraving.

Structural Feature "M": consisting of 174 longitudinal iron bolts (two per bay, for each arch), probably 1-in., and which extended longitudinally from one end of the bridge to the other, between the tops of the doubled king posts and just under the truss ties.

Structural Feature "N": consisting of 12 iron bars (two at each end of each arch), one and one half inches square, "secured in the bottom of each abutment, and passing up through the great body of masonry to the surface, and from thence to the top of the first king post of each range. . ." (from the 1814 Report).

Most of the remaining wooden structural members, specifically the floor beams, joists, and other bracing members, are not listed here as they are hard to reconcile with the 1814 Report and with the engraving, and they are not vital to the structural understanding of Wernwag's bridge. Furthermore, there are some mathematical miscalculations in the 1814 Report that make it difficult if not impossible to be sure about the lengths or sizes of the less important members. Despite these errors involving the sizes and weights of several members (by species of wood), it is evident that the bulk of the structural system was constructed of white pine, and it is likely that the flooring was made of yellow pine. Perhaps the lattice work under the floor was also made of yellow pine. It would be speculative to make assumptions beyond this point, although there was a very small quantity of oak used. The figures for the "Colossus" were listed as 10,993 ft^3 of white pine, 3,576 ft^3 of yellow pine, and 162 ft^3 of oak. Eastern white pine was the primary wood used for most building and bridge construction in the Delaware Valley in the late eighteenth century and well into the nineteenth century.

Elevation drawing, scaled ½ in. = 1 ft (reduced here), by Lee H. Nelson, 1987.

cont'd from page 27

been rather sizeable castings for their day (about 14 in. by 50 in.). While the "Colossus" was primarily an achievement of wooden arch construction, Wernwag should also be remembered as an early and extensive user of cast iron for bearing blocks, key plates, and wrought iron for a variety of bracing. Perhaps the most significant departure from traditional building technology was the fact that no mortise and tenon joints were used at the critical joints between the laminated ribs and the king posts. Instead, the king posts (doubled 6 in. by 12 in., spaced apart) were set on cast-iron "boxes" that straddled the full width of the two laminations of the ribs. There were holes in the middle of the cast-iron "boxes" for bolts because the king posts were secured to the arched ribs with an eye bolt on each side of the king posts, and the bolts extended down to embrace the ribs (see Fig. 20). The cast-iron boxes also appear to have been designed to provide a bearing surface for the cross bracing members. We assume that these castings were simply lagged to the top of the ribs rather than being dapped into the ribs, because that would have made it almost impossible to replace the ribs should that have been necessary. Such a possibility was one of Wernwag's claimed advantages for his design.

The king posts were of uneven length, shortest at the center of the bridge and becoming increasingly longer towards the abutments, but averaging about 15 ft in length and installed "normal" to the curve of the arch. There were 28 bays of king posts and bracing, each bay having a panel width of about 12 ft and an average height of 15 ft. The only exceptions to the "neither tenon nor mortice" joints, according to the completion report, were "a few to unite the king posts and truss ties." It is not clear whether that meant longitudinal ties or cross ties, but we believe it meant longitudinal ties.

By virtually eliminating the use of mortise and tenon joints, Wernwag intended to prevent (or at least reduce) the tendency for dry rot caused by wetting in concealed and inaccessible joints. In conventional wooden bridges with mortise and tenon joints, the disassembly and replacement of rotted interlocking truss members are virtually impossible. To prevent such problems would require a high level of mainte-nance of the roofing and siding, as seems to have been the case at the Market Street Bridge, where the tolls generated the necessary income, and where the mor-tise and tenon technology demanded such mainte-nance.

In connection with this and other bridges by Wernwag, there were numerous references to the importance of avoiding dry rot; and Wernwag seems to have devoted much attention to construction details that would minimize if not actually prevent dry rot. His principal techniques were to be sure that he was starting with sound timbers, to limit the size of any single member, to avoid mortise and tenon joints, to

minimize the contact of timbers, and to provide ample air circulation around timbers.

In the completion report, it was stated that:

> All the timber in this bridge has been slit [lengthwise] through the heart, so as to show any defect, and by being prevented from coming into contact, is secured from the dry rot.

At Wernwag's New Hope bridge, this claim was repeated and expanded as follows:

> The whole of the timber in this, as well as all Mr. Wernwag's bridges, is sawed through the heart to detect unsound pieces, and as the largest do not exceed six inches in thickness, they will readily season without danger of dry-rot. Whatever may be the strength required, the pieces of timber are not in-creased in dimensions, but in number, and carefully kept asunder by iron collars to admit the air.

The claim as to the maximum thickness of 6 in. is not strictly true for the "Colossus," because the list of timbers in the completion report includes cross ties that were 8 in. by 12 in. However, it is clear that the several members on this list described as being one foot square (king posts, truss ties, and braces) were actually double members measuring 6 in. by 12 in., and that they were listed as "1 foot square" for the purposes of calculating the quantity of "cubick feet." This is borne out by the mention that the ribs measured "3 feet 6 inches by 12 inches," when we know from other evidence that they consisted of six smaller ones.

Our assumption that the king posts and longitudi-nal truss ties (or top chord) were actually *doubled* 6 in. by 12 in. appears to be further supported by English engineer John Millington's 1839 description of the bridge, written just after the fire, "to preserve some record of its formation." His description, while not entirely clear, seems to suggest that half-laps were used to avoid mortise holes and that members were paired, or doubled. About the bridge he says that:

> . . .one of its peculiarities was that every large piece of timber was sawed lengthwise in two, in order to examine its heart, and see that no rotten, shakey, or bad timber was introduced into it. By this means mortice holes were avoided in the main timbers, as the tenons passed *between*, instead of through them, but were *let in* a sufficient distance on both sides to preserve them in their positions. This expedient like-wise proved a great preservative against dry rot, because all the timbers were so distant from each other as to permit a free circulation of air between them, except in the actual joints, and they were kept close by iron screw bolts . . . . the main arch consisted of three double rows of main timbers, laid three deep, or one above the other. Near to these, their correspond-ing halves were placed face to face, but with the tenons of what may be called the king posts between them, the whole being key joggled and bound together by wrought iron hoops [italics supplied].[50]

cont'd on page 34

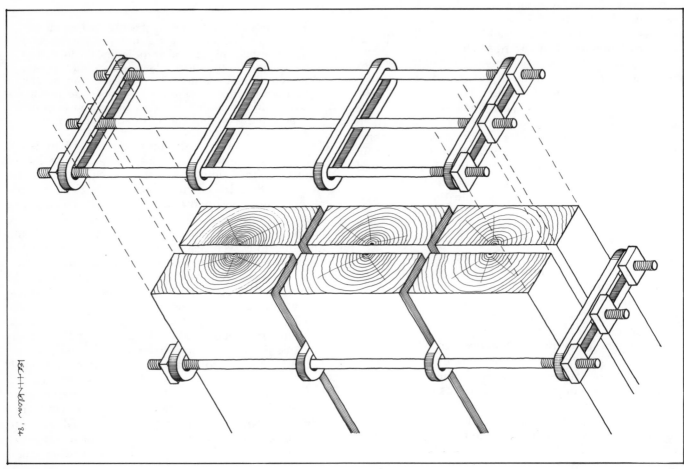

FIG. 19.
Perhaps one of the most unique aspects of Wernwag's design was the use of iron "links" and iron bolts to separate the six wooden members that comprised the primary arch, and which were intended to reduce dry rot by allowing air to circulate around them, while providing structural restraint and forcing them to act as one large composite member, e.g., a laminated arch of very large cross section (see Fig. 21 for evidence regarding the *shape* of the links and the cross section of the arch).

This construction technology seems to have been integrally related to Wernwag's approach to eliminate dry rot. As stated by Wernwag on the Gridley engraving: "The dry rot is entirely eliminated by the timber being sawed through the heart for the discovery of any defect & kept apart by iron links & screw bolts. . . ."

As John Millington stated in his *Elements of Civil Engineering* (Philadelphia and Richmond, 1839), ". . .every large piece of timber was sawed lengthwise in two, in order to examine its heart, and see that no rotten, shakey, or bad timber was introduced into it." This pairing and spacing of timbers that were sawn lengthwise was also utilized for the king posts and the truss ties. Wernwag utilized this concept in all of his later bridges, and it also appears that his competitors must not have regarded that separation of structural members as important to bridge design *per se,* as it was not utilized by later bridge builders. It seems unlikely that they were intimidated by the details of Wernwag's patent, when that was perceived as relating primarily to the configuration of the trussed arch.

Drawing by Lee H. Nelson, 1984.

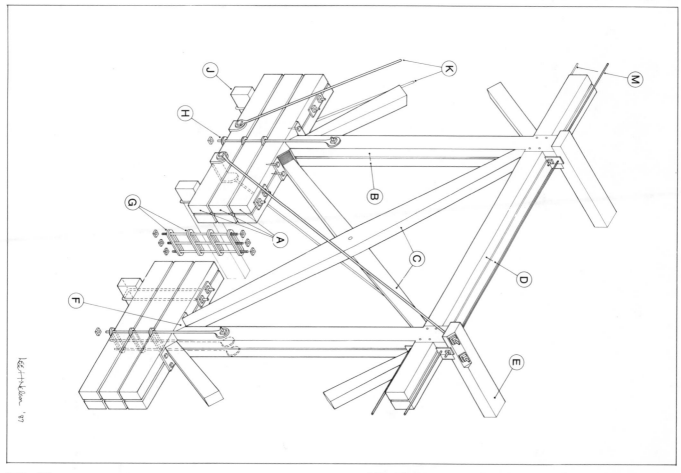

FIG. 20.

Isometric drawing of an arched rib, the trussing system, and the supplementary ironwork shown at mid-span with the major structural features related to the drawing with an alphabetical "key." See Fig. 18 for full descriptions of these structural features.

Structural Feature "A": Laminated ribs; at mid-span they were 6 in. × 12 in., white pine, arranged two timbers wide and three deep.

Structural Feature "B": Doubled king posts, 6 in. × 12 in., (nominal), white pine.

Structural Feature "C": Braces and counterbraces, 6 in. × 12 in., white pine.

Structural Feature "D": Truss ties, 6 in. × 12 in., white pine.

Structural Feature "E": Caps, 8 in. × 12 in., white pine.

Structural Feature "F": Cast-iron boxes.

Structural Feature "G": Wrought iron links and bolts.

Structural Feature "H": Iron eye bolts.

Structural Feature "J": Cross ties, 6 in. × 12 in., white pine, and iron stirrups.

Structural Feature "K": Iron hanging braces with stirrups and bolts.

Structural Feature "M": Longitudinal iron bolts.

Isometric drawing, scaled ¾ in. = 1 ft (reduced here), by Lee H. Nelson, 1987.

33

cont'd from page 31
The other construction aspect that should be mentioned is the very extensive use of wrought iron, in the form of iron tie rods, eye braces, stirrup irons, and iron abutment bars, all totaling some 40 tons, not counting the 11 tons of cast iron (for a detail of some iron members, especially delineated on the Gridley engraving, see Fig. 21). It is obvious that most of this was intended and was a part of the original design, based upon an itemized estimate of iron totaling over 46 tons. This estimate was prepared at a time when the design still called for five ribs instead of the three ribs as built. The estimate included quantities of iron bars of large sizes (1 in., 1¼ in., and 1½ in. square, and 1½ in., 1¾ in., and 2 in. diameter round bars), as well as long lengths (59 ft long for tying abutment piling, and many bars in lengths of 15 ft to 28 ft for ordinary bracing). In addition, it appears that some bars were added after the structural "scare," when the western abutment began to move, throwing the bridge out of alignment. Though it is difficult to determine the nature and extent of the remedial work, this writer is satisfied that the Gridley engraving does not show any of the remedial work, and thus the nature of that work remains obscure.

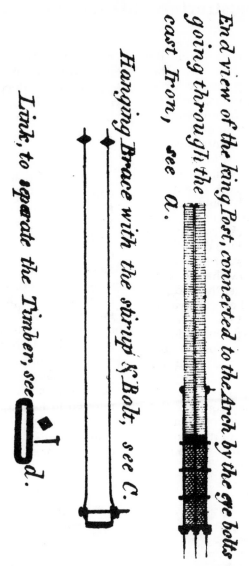

End view of the king Post, connected to the Arch by the eye bolts going through the cast Iron, see a.

Hanging Brace with the stirrup & Bolt, see C.

Link, to separate the Timber, see d.

FIG. 21.

Three enlarged details from the Gridley engraving (Fig. 16). These details have been reproduced here so that the two right-hand details can be seen "normal" to their use in the bridge.

One of the remarkable aspects of the Gridley engraving was the attention to detail, especially as it relates to the use of iron for structural purposes, including the very early use of cast-iron "boxes" for bearing the king posts and cross bracing, together with the use of iron hanging braces, iron stirrups, iron "links," iron tension rods, and the extensive use of iron bolts. Perhaps because this bridge made an unprecedented use of such iron, Wernwag may have wanted his broadside (the Gridley engraving) to "advertise" that fact. In any event, the broadside was unusual in its depiction of iron details as seen here:

Left-Hand Detail: One of the 579 "links" that were used in conjunction with bolts passing through the links to separate the "laminations" of the ribs (see Fig. 19) so that all of the wooden ribs were spaced apart by 1-in. to allow for the circulation of air and to help prevent dry rot.

Middle Detail: One of the "hanging braces" and "stirrups" that tied the primary arched ribs to the upper cross caps, and which were apparently intended to assure that one of the cross braces would remain in compression at all times. See Figs. 18 and 20, which show these hanging braces in their structural context.

Right-Hand Detail: This is a section showing the six "laminations" of the arched ribs, spaced apart with the iron links and secured to the paired king posts with long iron "eye bolts," which both embraced and separated the king posts. In other words, there are three "eye bolts" at each panel point (see Fig. 20, Structural Feature "H"). Note that the middle "eye bolt" *passed through* the cast-iron shoe at the juncture of the king posts and the ribs.

Illustration is extracted from a copy of the Gridley engraving from the writer's collection.

35

Superstructure: Statics Discussion

Actually, we probably know more about the physical aspects of the "Colossus" than any other bridge in early American engineering history. From the 1814 report published by the Managers, we know the size, species, and weight of all of the wooden members in the superstructure (including the 3-in. thick floor of the bridge). That same report gave the weight of the cast iron and the bar iron in the superstructure, with the combined weight of the wooden and iron members totaling 347 tons, 3 hundredweight, and 14 pounds, which comprised the basic dead load of the entire superstructure.[51] In addition to these "known" loads, we can estimate the other dead loads, which would include the covering of the bridge, consisting of the wall and roof framing, the siding and the shingling. We also know that 40 to 50 head of cattle were permitted to pass over the bridge at one time, thus constituting one kind of an "allowable" live load, although it is quite likely that heavier wagon loads were allowed.[52] To all of this we could add the estimated snow load, which would give us the total dead and live load conditions for the bridge (but not the wind loads).

We also have enough good dimensional information to establish the basic geometry and configuration of the superstructure, with its 28 bays of common width but varying height, consisting of the arches, king posts, bracing, and upper truss ties. By mathe-matically combining the geometrical and dimensional data with the various dead and live loads, it has been possible to conduct a computer analysis of the "Colossus" to determine the forces and stresses in the individual members, and to determine its overall structural "behavior." This, in turn, has enabled us to evaluate the structural rationality and efficacy of Wernwag's design.

The computer analysis was carried out in May and June of 1987, by Jon E. Morrison, P.E., Keast and Hood Co., Structural Engineers, Philadelphia, Penn. Mr. Morrison used the computer program called GT STRUDL, a sophisticated Georgia Institute of Technology program for assisting engineers in structural analysis and design decision making. Mr. Morrison's summary report, included here, is based upon a 53-page computer printout, and he describes the basic information and assumptions that were used in the computer "modeling." While this analysis was carried out by Mr. Morrison in collaboration with the writer, the overall direction for this analysis was provided by Nicholas Gianopulos, P.E., Keast and Hood Co. To better understand the summary report, the reader is referred to Figs. 16, 18, 19, and 20, but especially to Fig. 22 for graphic representation of the geometry, configuration, and forces within the various members of the "Colossus."

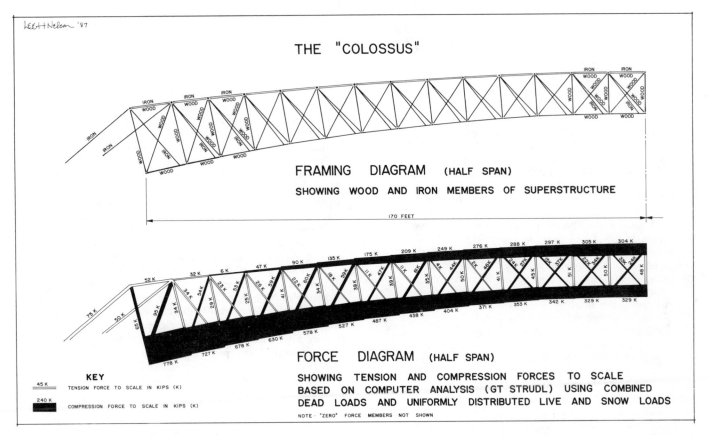

THE "COLOSSUS"

FRAMING DIAGRAM (HALF SPAN)

SHOWING WOOD AND IRON MEMBERS OF SUPERSTRUCTURE

170 FEET

FORCE DIAGRAM (HALF SPAN)

SHOWING TENSION AND COMPRESSION FORCES TO SCALE
BASED ON COMPUTER ANALYSIS (GT STRUDL) USING COMBINED
DEAD LOADS AND UNIFORMLY DISTRIBUTED LIVE AND SNOW LOADS

NOTE · "ZERO" FORCE MEMBERS NOT SHOWN

KEY

45 K — TENSION FORCE TO SCALE IN KIPS (K)

240 K — COMPRESSION FORCE TO SCALE IN KIPS (K)

FIG. 22.

The "Colossus" Superstructure. Framing diagram (above) and force diagram (below), both drawn to show half the total span, the other half being symmetrical about the center line.

Framing Diagram (above): This simplified diagram shows the basic configuration of the superstructure, as to the disposition of the wooden and iron members. See Figs. 18–20 for details of the various structural members.

Force Diagram (below): This half-span diagram shows the comparative tension and compression forces in the structural members, based upon the GT STRUDL computer analysis, using: (1) The known dead loads of the superstructure; (2) the estimated dead loads of the covering (walls and roof); and (3) the estimated live loads and snow loads, uniformly distributed.

For the purposes of this drawing, the various computer-derived forces (in kips) are drawn to a uniform scale using black for the compression forces and "open" paired lines for the tension forces.

The superstructure physical geometry, for the purposes of the computer "model," is based upon the Gridley engraving (Fig. 16); each of the three arches are assumed to be equally stiff, although the two outer arches were actually "pinched" in plan (see Figs. 15 and 23). The computer "model" also uses Wernwag's wood densities (e.g., 44 and 48 lb/ft³ for white pine and

yellow pine, respectively), which are considerably higher than today's figures, and we assume that the wood was wet and laden with resin when he weighed representative samples. Thus the computer analysis is a "worst case" scenario in several respects.

It should be noted from the computer analysis that there were a number of "zero" force members, which are therefore not shown on the force diagram. These "zero" force members include a number of the wooden counterbraces except for the four bays each side of center, the iron hanging braces in the same four bays, and most of the iron rods above the top chord (except in the two end bays).

Note that all of the king posts were in tension, except for the end king post. Thus the tension in most of the king posts was actually carried by the iron "eye" bolts (Feature "H" in Fig. 20). Also note how the compression forces in the top chord were gradually transferred into the primary arched ribs, by way of the wooden braces. Finally, note that the iron "stay" braces that extended from the end of the superstructure down to the abutment were helping to pick up some of the forces in the main arched ribs; in fact, those stay braces were the most highly stressed members in the bridge (see the "Summary Report on the Computer Analysis," by Jon E. Morrison, this book).

Drawings by Lee H. Nelson, 1987.

37

Summary Report on the Computer Analysis
of the "Colossus"
by Jon E. Morrison, P.E.

The loading for the computer "model" consisted of Wernwag's listing of all of the wooden structural members, their sizes, species, and weight, together with his figures for all of the iron in the bridge, totaling 694.3 kips. In addition, we have utilized estimates for the other dead loads, siding and roof, amounting to 183.6 kips and 265.2 kips, respectively. To this we have added your estimated maximum live loads, consisting of three wagons loaded with quarried stone, each weighing 20 kips; and, finally, we have added a snow load. The sum total of these dead and live loads is 1,528,314 pounds.

Please note that the load values shown in the computer output represent one-third of the total for the bridge, our assumption being that the three arches were equally stiff. After some discussions and preliminary analysis, our final model was run assuming that the wrought iron links and bolts [item G, Fig. 18] provided horizontal shear transfer between the members that made up the ribs, effectively forming a single composite or "laminated" flexural unit.

In addition, the timber braces and counterbraces [item C in Figs. 18 and 20] were only allowed to develop compressive forces, since they were not "tied" to either the ribs or the upper chords. The iron bars K, M, and N were assumed to only develop tensile forces. The timber king posts "B" were modeled as sustaining either compressive or tensile forces since they were tied at each end.

Referring to the force diagram [see Fig. 22], it can be seen that near the mid-span, the top and bottom chords are in compression, and that the iron and timber diagonals are very lightly stressed. This implies that this mid-span zone of the structure is acting in a true "arch-like" fashion. Nearer the abutments, at the quarter points, it is evident that a transition is occurring. The compressive force in the top chord is starting to feed into the timber diagonals and then into the bottom rib at the abutments; and the iron rod diagonals are starting to develop and assist in supporting the bottom rib, ultimately being balanced across the top of the first king post and into the back stay rod in a "cable-stayed" manner at the abutment.

The levels of deflection and stress at critical locations seem quite reasonable. The mid-span deflection under full loading is approximately 10 in. (1/400). The highest stress in the timber top chord at mid-span is around 2100 psi. This is high but not unreasonable for the level of applied load in the model, which includes the three wagons loaded with stone and the snow load, a "worst case" scenario.

The axial stress in the bottom chord rib at the abutments is approximately 1350 psi, which is quite reasonable. All but the end king posts are in tension, which is taken up by the eye bolts [item H in Figs. 18 and 20] with a maximum stress around 350 psi, which is very low. The compressive stress in the end king posts is around 450 psi. The maximum compressive stress in the timber diagonals is around 1300 psi, while the maximum tensile stress in the iron diagonals is approximately 23,000 psi.

It is interesting to note that the *only* top chord iron rods that are working are those in the first two panels near the abutments. The maximum stress in those members and the iron back stays at the abutments is around 33,000 psi. Again, these are reasonable values. Also note that the only places where both the braces and counterbraces are working (in compression) are in the four bays adjacent to the mid-span. In other words, most of the counterbraces are redundant, as are most of the top chord iron rods and a number of the diagonal iron braces [item K in Fig. 18].

Despite these redundancies, Wernwag's "Colossus" was a bridge that worked very well—an extraordinary piece of engineering design.

Wernwag's Design in Historical Perspective

With the computer analysis providing a modern evaluative background, it is interesting to take a fresh look at the historical context for the design and construction of the "Colossus," and to ask some questions about the circumstances surrounding the design of this bridge.

How was it possible for this remarkable structure to be built 175 years ago? While there are several possible answers to this question, they involve, at the very least: (1) Looking at the immediate engineering context for the "Colossus;" (2) looking at the general "tenor" of the era, which produced several important and unusual bridges; (3) a look at the role played by the client in this instance; and (4) trying to evaluate the engineering competency, experience (and persuasiveness) of Lewis Wernwag.

The technological state of the art in bridging leading up to the "Colossus" was really embodied in two bridges discussed earlier, namely the "Permanent" Bridge across the Schuylkill in Philadelphia, built 1801–1804 (see Figs. 8–11), and the Trenton Bridge over the Delaware, built 1804–1806 (see Figs. 12–13).

The "Permanent" Bridge was a major construction involving the best engineering talent available: William Weston from England, who designed the coffer dams for the piers, and Timothy Palmer, who designed and built the superstructure. It was successful by all standards, structurally, aesthetically, and financially; and it established the standard in the opening years of the nineteenth century for multiple-membered arched ribs of relatively long spans (150 ft, 195 ft and 150 ft). Having said that, we might ask whether the engineering empiricism that produced the "Permanent" Bridge was enough to produce an engineering superlative like the "Colossus"? The answer is undoubtedly, no. It is probably reasonable to assert that the principles utilized by Timothy Palmer in the "Permanent" Bridge could never have been extended to the scale of the "Colossus," partly because of the extensive use of dapped joints between the king posts and the ribs (thus weakening the ribs), and partly because of the bending introduced at critical points in the ribs, especially near the piers. In fact, Palmer is reported (by the President of the bridge company) to have said, after he examined the site and the intended piers for the "Permanent" Bridge, that he,

> pointedly reprobated the idea, of even a wooden arch extending farther than between the position of our intended piers, to wit, 187 feet. He had, at the *Piscataway* bridge, erected an arch of 244 feet; but he repeatedly declared that, whatever might be suggested by theorists, he would not advise, nor would he ever again attempt extending an arch, even to our distance, where such heavy transportation was constantly proceeding.[53]

In other words, Timothy Palmer, having built a 244-ft span (probably the longest in America at the time), and representing the state of the art in bridge-building technology, seemed to be acknowledging that he had overreached the practical limits of his engineering design approaches.

The next important bridge structure to be built was the five-arch span over the Delaware in Trenton, N.J., in 1804–1806, by Theodore Burr (see Figs. 12 and 13). This bridge is deserving of a serious historical study in its own right. It consisted of arches that were truly "laminated" with relatively light members bent to the curve; and while its "new" technology must have offered great promise for multiple-membered arched ribs, its longest span (200 ft) was still a long ways from the herculean scale of the "Colossus." In short, none of the bridges built in the first decade of the nineteenth century could be regarded as prototypical to the "Colossus," either in terms of design or in terms of their scale as clear spans.

Even if these bridges did not provide a direct design path to the "Colossus," they were still very interesting, and we wonder what was there about the times that produced such an array of innovative structures in the early nineteenth century?

First, the times were expansive and receptive to ambitious internal improvements, whether canals, dams, turnpikes, or bridges. In the year 1812 alone, five major bridges were commenced in Pennsylvania. Three of them were far more ambitious than the "Colossus" in terms of their cost, total length, and number of arches. These were the bridges in Pennsylvania, at Harrisburg, Northumberland, and Columbia. Collectively, they involved $750,000 of both private and state money (versus $80,000 of private money for the "Colossus"). In addition, their overall lengths were 2876 ft, 1825 ft, and 5690 ft, respectively, and their number of arches were 12, 8, and 53 arches, respectively (compared to a single arch for the "Colossus").[54]

Secondly, Wernwag benefitted from receptive and "ready" clients in the Managers of the Lancaster-Schuylkill Bridge Company. They had seen the success of the "Permanent" Bridge at Market Street, and they envisioned an equally remunerative venture at the site of the Upper Ferry, which served the turnpike to Lancaster. Furthermore the Managers had solicited proposals from bridge builders in the Philadelphia newspapers stating that the bridge was "not to have more than one pier (a single arch would be preferred). . ."[55] This is a very telling statement, because it indicates not only their receptiveness to a daring plan, but their hopes that it would be forthcoming, even if it far exceeded previous experience.

Although the Managers had received a single-arch proposal from Thomas Pope, it was clearly too fantastic for serious consideration; and the only other reasonable proposal was from Robert Mills, but it too was ordered "to lie upon the table." Subsequently, the Managers seemed to have been resigned to a two-arch bridge (probably by Mills), until Wernwag submitted "his plan," at which time the Managers quickly resolved that "Mr. Wernwag's plan of a Bridge with one arch of 330 feet chord be adopted."[56] Thus, the receptiveness of the Managers to a clearly untried and unprecedented long span arch seems undoubted.

If the times were right, and if the client was receptive, to what extent was Wernwag sufficiently experienced and persuasive to design, "sell," and construct such a bridge as the "Colossus"?

Given Wernwag's European background, and given his prior experience as a millwright and bridge builder in Philadelphia, there is every reason to believe that Wernwag had access to, and was aware of, the European testing regarding the compressive and tensile properties of structural materials that had occurred in the eighteenth century.[57] However, there is equal reason to believe that little of that knowledge about the strength of individual materials would have aided Wernwag in the design of a rather complex three-dimensional arched *system* like the "Colossus."

Not only had no one built on the scale of the "Colossus" before, but no one had devised a multiple-membered rib that worked so effectively as an arch, one which had the potential to be adapted to different spans by varying the number of multiple members or by varying the number of arches. As far as we know, even Wernwag had not built such a structure before, thus we look for other clues to explain his engineering competence and understanding.

One clue to Wernwag's approach to the design and construction is that he ascertained the weight of all of the wood in the superstructure by measuring the pieces after they were dressed and by weighing representative samples of the several species of wood to determine their weight. This suggests that Wernwag was well aware of the relationship between the loading and the useful working stress of eastern white pine wooden ribs that carried the majority of the loads in compression. Whether he understood the bending and shear forces is doubtful.

Another clue to his engineering understanding is the fact that he reduced his preliminary design from five ribs to three ribs, after the cornerstone laying but before erection of the superstructure. The *Minutes* of the Managers reveal that "Mr. Wernwag represented to the Board that three ribs instead of five, now exhibited in the model will be sufficient for all the purposes of permanence & stability of a good Bridge." The Managers agreed on condition that the two ribs should be added if it was found "necessary."[58] The three ribs were sufficient.

The above-mentioned scaling back of the bridge

from five to three ribs also suggests that Wernwag seems to have epitomized in all respects the confident and persuasive builder, with a daring but well thought out plan and ready to proceed with all of the complex logistics for such an undertaking. It appears that Wernwag was the only one to submit a proposal to the Managers who had actually built bridges. As a mill-wright and bridge builder, he had experience in framing timbers and he obviously understood the strength of materials.

Another fact that further suggests Wernwag's confidence was that his proposal had a built-in profit incentive for him to keep the cost down. And, subsequently, his success with the "Colossus" and the equally innovative New Hope Bridge resulted in the well-known broadside entitled "Wernwag's Bridges" that exuded professional confidence, and we might assume that his broadside was only an extension of the already existing entrepreneurial skills that may have been an essential part of his relationship with potential clients.

It might be said that by submitting a design for an unprecedented 330 ft span (its proposed length) Wernwag demonstrated his confidence (or arrogance), but the same could be said of Thomas Pope's submission (432 ft). The difference was that Pope's was *too* fantastic in that it depended upon upright wooden members, notched, so that each member was "nested" on the adjacent member, forming a giant cantilever; whereas Wernwag's concept was relatively a much simpler wooden arch, albeit a very long one.

To summarize this discussion, it appears that, despite some interesting and innovative bridges with multiple-membered arch ribs in the opening years of the nineteenth century, none of them seem to have been prototypical in design or scale to the "Colossus." And, though the times and the client were receptive to daring and inventive bridges, in the final analysis it was Lewis Wernwag who conceived the "Colossus," apparently with little outside inspiration, and who developed a remarkably rational system that worked very effectively, as is revealed by the computer analysis. If this is so, how did Wernwag arrive at such a clear engineering perception about the structural behavior of the "Colossus"?

It is this writer's speculation that Wernwag's engineering skills and confidence came about as a result of his use of *scale models*, and that he used models to accomplish two things: (1) To learn and understand the structural behavior of a given design, and (2) to demonstrate the effectiveness of his design and thus to help "sell" the design. In other words, perhaps the rationale and the persuasion came about as a result of the use of models.

We know that models were commonly utilized for demonstration and explicative purposes from the latter part of the eighteenth century, at least in and around Philadelphia. There is mention of several models having been used in connection with proposed designs for bridges over the Schuylkill, including one by the well-known architect/master builder Robert Smith in 1769; there are mentions of other models in the 1780s and 1790s, and there is a bridge model in the possession of the American Philosophical Society dating from 1786.[59]

Charles Willson Peale, in 1797, is known to have built a large model, loaded it with people, and made some comparative estimates regarding the weight that was sustained by the model as compared to a full-sized bridge. In February of 1812, in Philadelphia, Thomas Pope exhibited a model of his Flying Pendant Bridge, built at $\frac{3}{8}$ in. scale, and he charged $0.50 admission to view it.[60]

Other American bridge designers and builders are also known to have built bridge models, including Theodore Burr. While some models were built for patent purposes, it is possible that some were built to demonstrate the comparative strength of the design (as was done by Peale). There was specific mention later in the nineteenth century about experimental researches that were carried out utilizing models of various arch and trussing systems to determine the mathematical ratios between the weight of the model's framing system and that of the load borne by the model (without impairing the elasticity of the wood).[61]

Such comparisons were easy to make (although perhaps mathematically crude) and would have been logical extensions of the empirical bridge-building tradition. Although there is no proof to show that such was the case with Wernwag, he did in fact build a model.

These facts and circumstances suggest that models played a more important role in the design and promotion of early American bridge design than has been assumed in the past, and that the use of such models may have figured into Wernwag's design for the "Colossus." We know that his model was publicly displayed for admirers and critics alike (see the next section). Given the performance of the bridge as evidenced by the computer analysis, it is not unreasonable to assume that Wernwag had a good understanding of the structural dynamics at work in the "Colossus." Thus, it is tempting to believe that Wernwag's model may have played an important role in the development of this engineering superlative.

Even today, 175 years later, its 340-ft clear span is regarded as a daring and innovative concept. Considering the then available technology, the state of the art about statics, and the strength of materials, it is certainly not an exaggeration to say that Wernwag designed and built one of the most significant single-span wooden arched bridges in engineering history!

Wind Bracing

As was previously mentioned, it is clear from the record that Wernwag had built a model of the bridge "to illustrate to the public" his design for the bridge, and it is also clear that Thomas Pope, having seen the model, noted that it contained "valuable properties" belonging to his (Pope's) invention, as follows:

> namely, that the external perpendicular sides of said Model are built in the shape of two concave circles or arcs; their convex sides, of course, face each other, by which important shape, a more perfect resistence is furnished against the pressure that wind and tempest afford on the sides of a bridge. . . Therefore, the adoption of this shape. . .is a direct infringement on my patent right. . .and I. . .do require, that the said LEWIS VERNWAG [sic] forthwith alter, amend, or wholly destroy the aforesaid Bridge Model, so that it shall cease to partake of, or infringe on my patent right.[62]

This was part of a detailed newspaper advertisement by Thomas Pope specifically directed against Wernwag by virtue of Pope's Letters Patent for his "lever" bridge, which, he claimed, incorporated many "new and valuable properties . . . such as never before were realized in a structure of this kind in any country of the world."

Pope's advertisement appeared in mid-June, 1812, just when Wernwag was well underway with construction of the abutments and the securing of materials for building the superstructure. In fact, Wernwag was finalizing plans for that superstructure; and on July 2, he made a presentation to the Board of Managers regarding his proposed change from five to three ribs, mentioned before. We speculate that Wernwag had fine tuned his structural "calculations" and had assured himself that three ribs would be sufficient to carry all the live and dead loads. However, it also appears that, at the same Board meeting where Wernwag was seeking permission to modify the number of ribs, he also responded to Thomas Pope's public claims of patent infringement regarding the concave side walls of the bridge.

The handwritten draft of the *Minutes* for that meeting (July 2) includes two pencil sketches of the overall plan of the bridge, both of which showed the bridge wider at the abutments and narrower at the middle, but one of the sketches showed the transition being accomplished in a smooth and continuous curve, that is, having concave sides (like Pope's plan). The other sketch shows the plan "pinched" at the middle, and the transition being accomplished with straight line segments. It seems rather obvious from the pencil addenda to those *Minutes* that Wernwag proposed a technical circumvention of Pope's concave bridge plan by utilizing a simple "bend" in the ribs at the midpoint of the bridge, rather than Pope's concave shape. There is no further discussion of the matter, either by Wernwag or by Pope. However, upon due reflection, we would conclude that either approach would have introduced some difficult construction problems, in as much as the ribs would have to be curved in both the vertical and horizontal axes. Considering these complications to the constructional geometry, the concept of built-in bracing (regardless of the shape) must have been very important to Wernwag! The bridge would have been far easier to build without that shape.

This is an important aspect of the so-called engineering "advantages" built into the bridge and is one that was not noticed by the bridge's admirers, nor was that aspect shown in the several plans of the bridge that were published here and abroad, including the widely known and often reproduced (and incorrect) drawing by C. A. Busby and published in London in 1823. The only known view that accurately shows the "pinched" plan of Wernwag's bridge as built is the view which was authorized by the Managers. This is a comparatively rare print and had never appeared in print until very recently (see Fig. 15). In fact, this aspect of the plan was not known when the Smithsonian Institution's then-called Museum of History and Technology commissioned an otherwise accurate and handsome model of the bridge for its exhibit of American bridge structures. The efficacy of Wernwag's wind bracing must have been demonstrated to everyone's satisfaction when, in September of 1821, the roof and weatherboarding were "torn off" by a hurricaine, and it was noted by the Managers in the *Minutes* that:

> it appears necessary for the preservation of the said Bridge to have the Roof and weatherboarding replaced . . .

This occurred during a period when other bridges were known to have been totally blown down in storms and high winds. That such a large structure could sustain such heavy damage without lateral failure suggests that Wernwag's built-in wind bracing had other than theoretical merit.

The 1814 completion report described the three individual ribs as being 21 ft "asunder" at the abutments and 13 ft 1 in. apart at the middle of the bridge, so that the structural system of the bridge was over 42 ft wide at the ends and only about 26 ft at mid-span. Some later bridges had stay braces extending out from the abutments, but this writer is not aware of other builders that utilized built-in bracing as a part of the arch plan configuration.

The Abutments and the Structural "Defect"

Since the "Colossus" was a single-span bridge, it was not necessary to build the very expensive, dangerous, and time-consuming cofferdams for mid-river piers, as had been required for the "Permanent" Bridge across the Schuylkill at Market Street. In that instance, the cofferdams had required the best engineering talent available and had posed a more formidable obstacle than building the bridge's superstructure. Though the Managers of the Upper Ferry bridge site (as it was known early in the process) sought to avoid the problems of piers because of the hazards to navigation and the dangers to the bridge from ice and floods, they nevertheless were faced with the expense of building rather massive abutments, not only sufficiently massive to resist the enormous thrust of an unprecedented long span bridge, but of sufficient height to place the superstructure safely above flood waters and to allow the passage of canal shipping. Furthermore, the abutments had to accommodate the rather disparate site conditions between the east and west banks of the river.

The east side comprised the rocky site adjacent to the waterworks at the foot of the prominent outcropping known as Fairmount. The west bank of this Upper Ferry crossing was quite different, being an alluvial river bank with a thick muddy fill many feet above bedrock. To further complicate the construction, this river crossing was also subject to the river's tidal fluctuations (approximately 6 ft).

Building the eastern abutment was a matter of grading and quarrying stone from the infringing rocky site for the roadway and building a rather substantial stone abutment at the river's edge to receive the superstructure and the toll house.

Constructing the western abutment was another matter. This involved building a timber retaining wall (or "wharf") in the mud at the river's edge, but the overall task was made more difficult because it was necessary to build the western abutment on this alluvial bank to a height equal to that which was on the rocky eastern outcropping. Thus, Wernwag constructed an enormous grillage (or platform) of logs which he laid in the mud and upon which he could build the massive stone abutment. This grillage was configured so that the finished stone abutment measured 62 ft frontage on the river and extended 40 ft back to resist the thrust of the bridge's superstructure. Actually, this western abutment was extended with stone wing walls, between which he filled with earth to provide a gradual roadway up to the bridge from the lower level of the western river bank.

The grillage was designed as a rectangular network of logs (about three or four logs thick) with openings to receive pilings driven down to the bedrock. Some of the grillage and piling was also used to support a portion of the wing walls in addition to the abutment. The 1814 report stated that 599 piles were driven through the openings in the timber grillage, which contained 275,000 ft of timber! Despite this massive matrix of timbers, piling, and the stonework that comprised the western abutment, it subsequently shifted, at first manifested by a fissure in the north wing wall and then by movement which exerted a pressure on the upriver arched rib, thus raising that side of the bridge. For a time, this latter movement appeared to threaten the stability of the entire bridge (see Fig. 23).

When this worrisome event occurred, most observers assumed that the thrust of the arched bridge was causing the abutment to move. In reality, however, it appears that the angle of repose of the huge quantity of earthen fill behind the abutment was causing the northern wing wall to bulge outwards; and then the pressure of the earth (and perhaps the inclined plane of the piling) probably caused a portion of the stone abutment to rotate slightly against the upriver structural rib.[63] This problem was noticed soon after the bridge was completed (but not yet covered). The Managers watched as the fissure in the northern wing wall opened wider, but that concern gave way to genuine apprehension after the northernmost arch began to rise and throw the entire bridge roadway a foot out of level!

From May through August (1813), the Managers consulted with "experts" including Oliver Evans and Robert Mills in their efforts to avert disaster. Only weeks earlier, Oliver Evans, well-known Philadelphia millwright and engineer, had pronounced the bridge to be a

> Masterpiece of Architecture. . .probably the world cannot produce its equal. There is not a Mortice or tenant [sic] in the whole arch of 340 feet. It has a most Majestic appearance.[64]

On June 5, 1813, Oliver Evans examined the western abutment, and he had initially and mistakenly assumed that

> it was giving way to the pressure of the arch which is 340 feet span, but I was mistaken, the abutment had raised the arch. I conceived the idea that if it were possible to apply the weight of the arch to draw the abutments inwards with a force to balance the pressure outwards that by therfor [sic] means only the bridge could be saved. I soon discovered that a segment of semi circle inverted would if applied to bear a part of the weight of the Bridge and hitched to the abutment at each end would produce the desired effect. That this inverted arch may be made of wood planks bolted together. If the inverted arch be a segment of a greater circle than the rising arch of the bridge, the force to draw the abutments together will overbalance that of the raising arch [and] push them out.[65]

cont'd on page 45

43

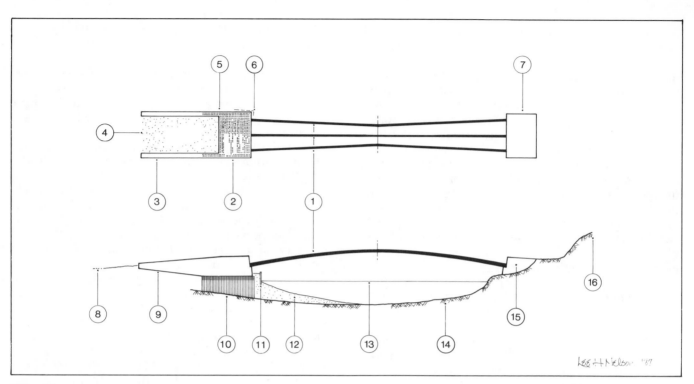

FIG. 23.

The "COLOSSUS." Diagrammatic plan (above) and cross section (below) showing the primary structural aspects of the bridge with its three ribs, the abutments, and the site conditions.

The Plan (above): Item 1: Three "laminated" ribs, each composed of six members (see Fig. 19). The two outer ribs are splayed, in plan, to provide built-in wind bracing. The ribs are 13 ft 1 in. apart at the center and 21 ft apart at the abutments. *Item 2:* Western abutment, 62 ft by 40 ft, built of stone on a grillage of timbers and piling, which encompasses the entire abutment and a portion of the wing walls. The "dots" on this plan suggest the 599 piles used for that purpose. *Item 3:* Wing walls, of stone, approximate length 100 ft. *Item 4:* Earthen fill between the wing walls, which was removed and replaced with stone after the abutment commenced to move. *Item 5:* This is the point where a "fissure" occurred (shortly after removal of the scaffolding from the superstructure) between the north wing wall and the western abutment, breaking all of the timber "ties" between the two rows of piling that supported those portions of the wing walls. The subterranean timber "ties" were replaced with several "chains" of 1¼ in. iron bars, but they too failed, breaking, as the north wing wall continued to move. *Item 6:* This marks the point where the western abutment rotated slightly, that is, it moved inward, compressing the northernmost rib and raising it about a foot and thus causing the entire roadway to be thrown out of level. Considering the geometry of the

bridge, it is clear that any slight movement of the abutment would cause a considerable raising of the arch. *Item 7:* Eastern abutment, 60 ft by 40 ft, built of stone on bedrock.

Cross Section (looking North): Item 1: The "laminated" ribs are seen here in section, being 3 ft deep at the center and 4 ft deep at the abutments, with a clear span of 340 ft, and a rise of 19 ft 11 in. from the point of springing. The roadway is 14 ft above the springing of the ribs at the abutments. *Item 8:* Approximate grade of the roadway, built of fill, to provide access to the western abutment. The access levels were determined by the eastern abutment (on stone). *Item 9:* Stone wing wall. *Item 10:* Grillage of logs (containing 275,000 "feet of timber") through which the 599 piles were driven to bedrock. The piles reportedly ranged in length from 15 ft to 30 ft, due to the inclined plane of bedrock. *Item 11:* Wooden "wharf" built to contain the mud fill around the abutment. *Item 12:* Natural mud overlying the rock river bottom. *Item 13:* "Common low water" line, approximately 18 ft below the springing of the arched ribs and 38 ft below the ribs at the center. There is about a 6-ft variation between low and high water at this point. *Item 14:* Approximate line of the bedrock, based (in part) upon modern river borings, taken somewhat further south. *Item 15:* Eastern abutment, of stone, built on bedrock, partially quarried away for that purpose. *Item 16:* The lower slopes of "Fairmount."

Drawing by Lee H. Nelson, 1987.

cont'd from page 43

It is not clear how this remedy was intended to work, although there is a suggestion in the report to the stockholders that Evan's suggestion was tried along with several other attempted remedies. On July 1, Wernwag was authorized by the Board

> to put a Chain along the upper Rib from king post to king post in such manner as he shall think proper Provided that if the Chain should in the opinion of the Board not answer their expectations he is to pay the expense.

On July 8, Wernwag made a "proposition" to the Board for "removing the defect in the Bridge," and while the details of that contract were not stated, they can be partially deduced from later events; but before that work could be effected, the Board of Managers decided (out of desperation) to report to the stockholders that the "unpleasant" situation was possibly beyond their control.

The July 31, 1813, "Statement" to the stockholders is especially interesting, for it describes the on-and-off progress of the "defect," as well as the "experiments" (unsuccessful at that point) for dealing with the problem:

> Soon after the Scaffolding was removed and the Managers were exalting in their complete Success in presenting to the publick a beautiful Arch of a chord nearly One hundred feet longer than any in Europe or America they perceived a small fissure in the North Wing Wall. Mr. Wernwag was immediately called upon to view it. It was attributed by him to the breaking of the ties of the Substructure which were large pieces of timber placed cross-ways to bind the two rows of piles upon which the wing-walls were built, and therefore supposed that the piling of the North Wall being no longer bound by these ties had inclined upstream, cracking the Stone work by its motion. To stop this he had several chains of 1-1/4 Inch iron bars stretched from Wall to Wall and drawn tight by Screws on the outside of each But these were unfortunately incompetent to the object; the North wall continued to incline outwards breaking the chains as it moved. The Managers then directed a quantity of Earth to be placed on the out side, along the defective Wall; and altho' many Carts and a Scow were constantly employed, and an immense body of Earth abutted against it, such was the Mass of Mud below, which the earth forced from its bed, before it could consolidate itself, that the motion of the wall was not stopt for many Weeks.
>
> The Managers distressed as they were at this event still had the satisfaction of finding that the abutment remained firm and that the Bridge exhibited a degree & beauty seldom surpassed in hydraulick architecture.
>
> Four months elapsed after the removal of the scaffolding from under the superstructure, without any other symptom of weakness appearing, than that before alluded to in the North Wing Wall, and this gave the Managers no cause of serious alarm, since its motion was nearly stopt, and not being materially connected with the mass of Masonry on the abutment, it was never supposed, in the worst event capable of

> doing it any injury. But about the beginning of May an inequality was perceived on the flooring of the bridge. This was instantly attended to, and when examined and leveled it was found that the upper and middle ribs had bent in the center and raised the North Section nearly a foot higher than the South. This was as unexpected as alarming. After a strict examination it was supposed to proceed from the pressure of the abutment upon the Superstructure either by the piles upon which the Masonry is built inclining their tops inward, or slipping below upon the Shelving Rock: Which ever of these causes may have produced the unhappy effect now visible on the bridge a remedy was instantly attempted which altho' not wholly effectual as yet had undoubtedly checked the evil. 7 or 800 perches of large stones were heaped directly in front, upon a vast mass of earth previously laid within the wharf. The Eastern motion of this abutment working as it does against the colossal weight of the Superstructure is as extraordinary as unexpected, since the whole attention of the builder was directed to prevent a contrary operation; and two strong arches were turned against the frame upon which the wing walls are built to stay the abutments from the weight and pressure of the superstructure.
>
> The Board of Managers have viewed this evil with a most anxious eye and deceived by its inconstant action some times stopping for two or three weeks, and still in hope of its ceasing its pressure altogether they have delayed asking a Meeting of the Stock-holders until this moment when they have deemed it necessary to resort to you for advice.
>
> Mr. Wernwag is, at his own proper cost removing the earth from behind the abutment, with an expectation that the weight of the Bridge upon the wall will bring the piles to their original bearing: In this experiment the Managers have no very great confidence. It was due however to Mr. Wernwag to suffer him to make the trial.[66]

Notwithstanding the "unpleasant" situation regarding the stability of the bridge, the tolls since its completion had exceeded $1200 and were expected to triple; thus the stockholders perceived that future profits would justify major expenses to correct the "defect." They authorized the Board to carry out one or more remedial actions, including shortening the compressed ribs or dismantling the superstructure and rebuilding the abutments and walls, and pay the costs thereof.

Shortly thereafter, several actions were taken, in part by Wernwag and partly at the suggestion of Joseph Johnson who was Wernwag's foreman and later collaborator on other bridges, including the multispan bowstring trusses across the Delaware at New Hope, then under construction.

A wide range of remedial efforts were proposed, a few of which were actually undertaken, but it is hard to sort them out. The known efforts included the piling of 700 to 800 perches of stone against the abutment to prevent further movement. In addition, earth was removed from behind the abutment and replaced with stone to reduce the lateral pressure of wet earth. The Board variously considered other steps, previously mentioned, but the exact nature of these remains vague.

In August, an order was given to "put in the Wooden braces between the King posts and a sufficient number of Iron chains under the direction of Mr. Johnson." The reference to "wooden Braces" is very unclear. They could have been new bracing and chains that were installed *between* the arches, that is, on the cross-axes; but it is also possible that they were the cross-braces for the superstructure that were being reinstalled after being experimentally removed (see the following for further discussion).

On September 23, 1813, the Board resolved that Johnson be presented with $200 "as a mark of gratitude and compensation . . . for his diligence and intelligence in remedying an unfortunate defect in the Superstructure. . .and of his uniform attention to their interest during the time of his attendance thereto."

It is this writer's opinion that little, if any, of the corrective work could have brought the abutment back into place. Removal of the earth from behind the abutment might have lessened the problem and, in the final analysis, may have been the reason the movement finally stopped; but in no event could the movement of the abutment be *reversed*, so Wernwag did the only thing that could "correct" the problem at the abutment, and he did this by cutting away the stone at the bearing point of the compressed upriver rib, thus allowing the rib to settle into its normal position, and bringing the roadway back into a level alignment! How he accomplished cutting away the stone when it was fully loaded remains a mystery. The solution was as incredible as it is vague.[67]

With the problems apparently remedied, the Managers authorized Robert Mills to proceed with his contract for enclosing and roofing the bridge. By the end of November, the covering was nearly completed and painted; and by year's end the sheetiron roofing had been installed on the circular roof of the toll house.

On January 6, 1814, the Managers reported to the Stockholders, that:

to the best of their belief the Bridge is perfectly solid in all its parts. The defect in the upper and middle ribs, which gave some uneasiness [!] last Summer, is now remedied, and the inward pressure of the western abutment stopt.

The final report of the Board of Managers was published on March 3, 1814.

Admittedly, these attempts to sort out all the remedial work that was carried out (in the summer and fall of 1813) are inconclusive; but let us now speculate on an even more important engineering matter of interest: does the Gridley engraving (see Fig. 16) show the bridge *before* or *after* the remedial work was carried out? While we cannot provide an unequivocal answer, we have little doubt that the Gridley engraving shows the bridge *before* the corrective work, for several reasons.

The first "state" of the Gridley engraving could have been made as early as (but not before) April of 1813, because that was when Wernwag and Johnson *started* the bridge across the Delaware River at New Hope; and the text of the engraving notes that the bridge was "erecting at NEWHOPE, 1813." This was only a month before the "Colossus" developed the problems that required remedial work several months later. It would appear that Wernwag rushed this promotional engraving into print before either bridge was finished because the second "state" of the Gridley engraving included additional (and corrected) text to reflect the completion of *both* bridges, probably after September, 1814, when carriages first crossed the "Newhope." However, the delineation of the structural details remained the same in both "states" of the engraving, despite changes to the text.

While the first problems with the "Colossus" started in May, 1813, only a month after the New Hope bridge was started, and the problems were essentially solved by August, it is barely conceivable that corrective work could have been shown on the first version of the engraving, assuming it was not published until that latter time, a technical possibility. This would allow the inclusion of braces or *counterbraces* "between the King posts," which had been ordered to be "put in" during August (see above). But there can be virtually no doubt that Wernwag designed the bridge to have *both* braces and counterbraces (rather than having counterbraces installed as corrective work, for example), primarily because of the way the diagonal braces were part of a consistent system of *paired* members, including the paired arches, king posts, and truss ties, all designed to keep the members small and replaceable. Furthermore, the cast iron "boxes" were a part of this paired system. It seems highly unlikely that Wernwag would have paired everything except the braces. Thus we feel that the engraving is an accurate depiction of the bridge as it was designed and built, except for the remedial work which remains vague; and it is further unlikely that Wernwag would have wanted to introduce anything that might have been construed as makeshift into his promotional broadside.

Conclusion

While this book has attempted to focus on the engineering and technological aspects of the "Colossus," those aspects should be seen in the context of the then contemporary American bridge design (and construction) state of the art. If it is possible to characterize the context of the "Colossus," the last decade of the eighteenth century marked the growing need for bridges and the willingness to invest in such enterprises, coupled with a rather dramatic and rather ingenious experimentation in statics by American bridge builders.

Certainly the leading bridge builder of the 1790s (and the first few years of the 1800s) was Timothy Palmer, who built a number of impressive bridges in the mid-Atlantic and Northeast states, culminating with the Schuylkill River bridge at Philadelphia and the Delaware River bridge at Easton, which were finished in 1805 and 1806, respectively. Though they were not as daring as the "stupendous arc" (Pope's words) of the 244-ft Piscataqua River bridge by Palmer, near Portsmouth, New Hampshire, the Philadelphia and Easton bridges represented the American state of the art in bridge building (and, incidentally, we know much more about them). Both were triple arched bridges (actually trussed arches), with the spans configured at 150 + 195 + 150 ft and 3 × 163 ft, respectively. Both were long-lived and attracted considerable attention from the engineering community.

The Easton bridge had the further advantage of being a through truss with a flat roadway. Both bridges were major construction projects, with much of the effort going into the piers. There had been little prior American experience in the building of cofferdams for the construction of masonry piers below water level. Despite these impressive efforts, Palmer's bridges would be considered wasteful by later standards, involving very large members, with a lot of custom fabrication for shoulders and haunches, and extensive use of mortise and tenon joints. The obviation of the latter practice was the focus of Palmer's successors, simply because it was wasteful and effectively limited the lengths of bridges that could be built with conventional trussing practices.

Only several years prior to the end of Palmer's era, Charles Willson Peale published (and patented) his design for a laminated arch bridge that involved smaller member sizes with a reduction in the number of mortise and tenon joints. That arch was truly laminated, the 2-in. thick members being laminated flatways and held together with locust treenails. Although historians have treated Peale's idea as an impractical intellectual diversion, his design did address contemporary concerns such as navigational obstructions and the sheer scale and complexity of trusswork.

Because of Peale's publication, his patent, the display of his model in his Philadelphia museum, and his contacts abroad—surely his ideas had an impact upon the American bridge-building scene, although this writer cannot document this assertion. Nevertheless, the next phase of experimental and even more daring bridge design during the next decade involved the use of "laminated" arches, both in the use of "bent" arches and laminated arches "cut" to the curve.

While Theodore Burr's later arch-reinforced *trusses* became the "standard" for years to come, his laminated *arched* bridge at Trenton was an anomaly and does not fit the structural concept normally associated with the so-called "Burr Truss," that is, a truss with a superimposed arch. During this period, Burr built even more daring spans, averaging about 200 ft and culminating in the McCall's Ferry bridge with the incredible span of 360 ft, but which had a very short life. It was taken out by ice in March, 1818, and little is known about the details of its construction.

However, the Trenton Bridge may be said to be the most advanced of Burr's bridges (the award for his most bizarre design must go to the Mohawk River bridge at Schenectady, New York, in 1807–1808). The Trenton bridge involved five spans of true laminated arches with a suspended roadway that was flat. The spans were 160 + 180 + 200 + 180 + 160 ft.

Burr must be considered the most experimental and innovative builder of the first decade of the nineteenth century (despite his loose and promissory business practices). Burr eliminated the mortise and tenon joints in the Trenton bridge, and he also eliminated the heavy members that characterized the earlier bridges. His laminated arches consisted of 4 × 12s laid flatways, they appear to have been conservatively designed, and the bridge was long-lived. Burr set the stage for Wernwag by building important bridges on a daring scale with innovative structural concepts that worked!

Finally, it must be admitted that the times must have been right for Lewis Wernwag, and, while we will never know the extent of the influence of his predecessors, Wernwag appears to have been a builder of considerable persuasion during a period that combined great economic optimism and a national need for internal improvements.

Wernwag's contributions to the art of bridge building were many. Although his first several bridges were modest in scale, his "Colossus" was his first and his greatest bridge utilizing the principles of multiple-member arch construction, which, while not conforming to today's definition of a laminated arch, it nevertheless was an innovative and superlative design that captured the imagination of both the romantic and the technological minds of the day, the impact of which extended well past the time of its construction.

In fact, no other American bridge commanded such admiration both in the United States and abroad (see Figs. 24–28).

Probably because of the problems experienced with one of the abutments soon after erection, and possibly because no other site demanded such a span, Wernwag never again attempted a span on such a scale. In some instances, Wernwag simply functioned as contractor rather than as designer (for example, the several bridges that he built for the Baltimore and Ohio Railroad). However, several of his less daring bridges built upon the same principles as the "Colossus" were very long-lived, one of which (the Camp Nelson bridge in Kentucky, 1838) was subjected to a modern stress analysis in 1927. The results of that analysis (published in *Engineering News-Record*, February 4, 1928, pp. 234–235) revealed that the fiber stress was well within today's acceptable limits and demonstrated the basic understanding that Wernwag had in his use of "laminated" wooden arches. While it is interesting to speculate about the various structural influences upon Wernwag, including those of Delorme, Gilly, Wiebeking, Peale, Burr, and Pope, it should not be ruled out that Wernwag might have seen the massively arched Wettingen bridge (which was "laminated," by our definition) built by the Grubenmann brothers in 1764–1766. While this Swiss bridge was well known through publications, Wernwag conceiveably saw it when he made his way from relatively nearby Reutlingen in Württemberg to America in the 1780s. In addition to the Wettingen bridge, it is also possible that Wernwag saw some of the other examples of Eastern Swiss bridge carpentry by the Grubenmann brothers, which included several "laminated" arches built before Wernwag left his native Württemburg.[68] However, none of these influences has been established, and Wernwag's superlative construction of the "Colossus" stands as one of the preeminent structures of American bridge construction without any attributable prototype.

Certainly some of the credit must go to the Lancaster Schuylkill Bridge Company for envisioning a bridge wherein "a single arch would be preferred." Nevertheless, Wernwag's engineering confidence must be considered the prime factor in the success of the "Colossus." This confidence was expressed in a number of ways, including his decision to reduce the number of arches from five to three ribs, which he represented to the Managers "will be sufficient for all the purposes of permanence & stability of a good bridge." Wernwag's contributions go beyond the sheer daring to the more interesting use of "laminated" timbers, at least in the technical sense if not in the conventional sense, in as much as they were multiple large-scale members cut to the curve and spaced apart with iron links and bolts and designed to be replaceable when needed (probably impractical). Not only did these "laminated" rib members increase in depth from mid-span towards the abutments, but the panels themselves increased in height towards the abutments. Furthermore, Wernwag cleverly utilized (even infringed upon) Thomas Pope's idea for built-in wind bracing by designing the outer ribs to splay out from the center toward the abutments. Finally, he designed the "Colossus" to minimize the use of mortise and tenon joints, and he used an extraordinary amount of wrought iron and cast iron for both bearing members and tension members.

In spite of the confusion about the authorship of the "Colossus" (Mills versus Wernwag), and despite confusion about certain aspects of its construction (witness the Busby drawing of the bridge; see Fig. 24), there is sufficient primary documentation to establish the all important and unique aspects of its construction. Certainly the aesthetic aspects of a bridge that enhanced the landscape were a combination of the daring span, the architectural treatment of the portals, and the enframed window openings in the sidewalls, which were designed by Robert Mills. But the primary basis for its admiration in the United States and abroad was clearly due to the invention of Lewis Wernwag, self-styled *Pontifex Maximus*![69]

THE UPPER SCHUYLKILL BRIDGE at PHILADELPHIA. from the S.E.

FIG. 24.
This structural delineation of the "Colossus" was engraved from a drawing by C.A. Busby and published by Taylor in London, 1823. This English version was incorrect in several important respects. The trussed framing system is notably wrong in that it does not include the counterbracing, the number of bays is incorrect (30 instead of 28), and the plan is incorrect in that it shows parallel arches rather than arches that flare out from the center toward the abutments. Considering that Busby came to America and appears to have carefully recorded the Trenton Bridge, it is curious that his drawings of the "Colossus" were so structurally careless; and if he visited it (and the Permanent Bridge), most of his attention seems to have been focused upon the exterior architectural aspects that were rendered for their picturesque qualities.

This view was probably responsible for several subsequent incorrect structural delineations of the bridge, including the one republished from the German work by J.G. Heck, *Iconographic Encyclopaedia of Science, Literature, and Art,* translated and edited by Spencer F. Baird, New York, 1851. Such an incorrect version was also published in Appleton's *Cyclopaedia of Applied Mechanics,* New York, 1880.

Illustration courtesy of the Rare Book and Manuscript Library, Butler Library, Columbia University.

FIG. 25.

Upon its completion in 1814, Wernwag's "Colossus" (officially called the Lancaster-Schuylkill Bridge, but more commonly known as the Upper Ferry Bridge and later as the Fairmount Bridge) quickly acquired status as an engineering marvel, was widely admired for its beauty and setting, and became a tourist attraction, in part because of its proximity to the Fairmount Waterworks and pleasure gardens. In this view, the waterworks are not seen, being behind the base of Fairmount at the extreme right (see Fig. 28). At the left of this view is the Upper Ferry House, also known as Harding's Hotel; and in the distance (above the bridge) is Henry Pratt's country house known as "Lemon Hill," which still stands as one of the historic houses of Fairmount Park. Pratt was the second president of the bridge company (elected in January of 1813).

This bridge was pictured by many American illustrators, starting with the well-known artist Thomas Birch (1779–1851), whose handsome painting was used by several printmakers, including the one seen here. In addition to numerous engravings, aquatints, woodcuts, and lithographs, the bridge was pictured on Staffordshire china (Fig. 26). This same view of the bridge also appeared on printed *toile* (cloth hankerchiefs), which attests to the interest in the subject since that medium was more often reserved for historical or moralistic subjects.

There were also many European views of the bridge. Some of these were plagiarized from Birch's view, including a well-known aquatint by Carl Fredrik Akrel, from a drawing published by Axel Leonhard Klinckowström, published in Stockholm, in 1824. There were numerous other views executed by artists and printmakers from Russia, Germany, France, and England.

The original oil painting by Thomas Birch is owned by the Historical Society of Pennsylvania and was the basis for this illustration, engraved by Jacob J. Plocher. The caption which accompanies the Plocher engraving reads as follows: "THE UPPER FERRY BRIDGE Over the River Schuylkill near Morris street in the County of Philadelphia. Chord of Arch 340 feet—whole extent of Bridge 400 feet—rise of Arch 20 feet—elevation above water 30 feet—the Span is greater by 98 feet than that of any other Bridge known—the construction is in general new—the principle invented by Lewis Wernwag who was assisted in the execution by Joseph Johnson—general design by Robert Mills. Architect."

Courtesy of the Atwater Kent Museum, Philadelphia.

FIG. 26.
Thomas Birch's view as engraved by Jacob Plocher (Fig. 25) was the source for the design on this English Staffordshire china plate, made by Joseph Stubbs. It was also used on platters, trays, tureens, pitchers, and vegetable dishes. At least three other views of the bridge were used for Staffordshire dinnerware by other makers.

Courtesy of the Atwater Kent Museum, Philadelphia.

FIG. 27.
This watercolor by David J. Kennedy is dated 1836 and carries the hand-written notation: "The Upper Ferry Bridge over the Schuylkill at Fairmount burned on September 1st 1838. Sketched from the balcony of 2nd Story of Upper Ferry House [Harding's Hotel] in 1836." The veranda of this hotel must have provided a grand vantage point when the bridge burned.

With the Fairmount Waterworks and the reservoir in the background, no other known view so successfully captured the sweeping curve of Wernwag's "Colossus." It could have been this dramatic prospect that prompted actress Fanny Kemble's romantic similitude of 1832, ". . .at a little distance, it looks like a scarf, rounded by the wind, flung over the river."

Courtesy of the Historical Society of Pennsylvania.

FIG. 28.

This painting of the Fairmount Waterworks best shows the man-made and natural setting of Wernwag's bridge, with the reservoir above, the dam (at the left), the Schuylkill River (tidal to the dam), and in the right foreground is seen one of the locks of the Schuylkill Navigation Company canal, which brought canal boats around the dam. A hint of the pleasure gardens is seen with the plantings and the stairways to the "Mount." This was one of the best-known scenes of early nineteenth century Philadelphia, and it became the subject of numerous engravings. By this time, the bridge was commonly known as the Fairmount Bridge, although some of the engravings (made just after the fire) label it as "the late Wernwag's bridge," and though he had long left the scene and was living in Harpers Ferry in Virginia, his fame and name were inextricably associated with this engineering superlative.

Illustration courtesy of The Athenaeum of Philadelphia.

Footnotes

1. There are several and varied works that generally describe the pervasive and long-standing need for bridging the numerous rivers and coastal waterways that frequently separated American seaboard cities from their hinterlands. These include such works as Carl W. Condit, *American Building: Materials and Techniques from the First Colonial Settlements to the Present* (Chicago, 1968, pp. 22–25), and the several books by Richard Sanders Allen, *Covered Bridges of the Northeast* and *Covered Bridges of the Middle Atlantic States* (Brattleboro, Vermont, 1957 and 1959, respectively), and the dissertation by George Danko, entitled "The Evolution of the Simple Truss Bridge 1790 to 1850: From Empiricism to Scientific Construction" (University of Pennsylvania, 1979, pp. 2–9). For more localized (Pennsylvania) references to citizens' petitions for bridges, official reports of committees established to consider appropriate "scites" or plans for such bridges (including the involvement of citizens like Benjamin Franklin), discussions about the merits of bridges over ferries, or discussions about particular proposed crossings, all ranging from the 1750s to the end of the century, see the *Pennsylvania Archives*, Eighth Series, vol. IV, 3407–3408, 3418, 3440 (1750s): vol. VII, 5711–5713, 5981, 6035, 6335–6336 (1760s); and vol. VIII, 7179–7181, 7187, 7204–7205 (1770s, but, actually, there are numerous other references from the 1750s on). For a variety of other relevant sources, see *Columbian Magazine* (January, 1787), *Extracts from the Diary of Jacob Hiltzheimer. . .* (Philadelphia, 1893), *The Pennsylvania Packet. . .* (Nov. 2, 1790, p. 3), or the Notes of James Todd on the Arguments of Edward Tilghman, Miers Fisher, and William Lewis on the Act to incorporate the Subscribers to the plan for erecting a permanent Bridge over the River Schuylkill—March 6 and 7th 1788, Independence National Historical Park Collection, Catalog No. 628. For references to bridges further afield, see the *Minute Book*, Library Company of Philadelphia, vol. IV, p. 18 (5 February 1795) for mention of a model and a drawing of Palmer's 244-ft single span over the Piscataqua River in New Hampshire. Also see a 1795 letter from Joseph Sansom (a Philadelphia merchant) addressed from Boston to his parents, in which he says "the peculiar glory of this Country are the numerous and immense wooden bridges of different construction." He mentions a 180-ft "beautiful arch" span over the Merrimack, "which being painted white glitters, like a faery vision . . . " as well as a three-arch bridge at Haverhill (with arches from 150 to 180 ft); "But the chef d'ouvre is a single arch of 244 feet striding like a Mammoth over the channel of Piscataway [sic]." He further mentions a proposal to build a 500-ft span over the Kennebec, in Maine. See the "Morris-Sansom Papers," Quaker Collection, Haverford College. In November, 1786, John Sellers "produced a model" of a bridge to be erected over the Schuylkill in Philadelphia at a meeting of the Society of Agriculture at Carpenter's Hall, which was mentioned in *Extracts from the Diary of Jacob Hiltzheimer. . .* (Philadelphia, 1893, p. 103). The American Philosophical Society owns a wooden model of an open, arched bridge, which was presented by John Jones, Sussex County, Delaware, in 1786. For the long-span design of 1796 over the Schuylkill by Godofroi Du Jareau, see the original watercolor, which is also owned by the American Philosophical Society in Philadelphia.

2. Charles Willson Peale, *An Essay on Building Wooden Bridges* (Philadelphia, 1797). This pamphlet has 14 pages of text and six engraved plates, which include a plan, elevation, a cross section showing the mode of laminations with their treenails, a detail of the railings with their stanchions and ships knees, two plates showing rather diagrammatically the modes of assembly upon the river banks and erection from floats in the river without the use of scaffolding, and a plate showing an arch "made flatter by additions at the butments [sic]."

3. The writer is indebted to Sidney Hart, National Portrait Gallery, Smithsonian Institution, Washington, D.C., for sharing translations of the evaluation of Peale's design carried out by the French Academy of Sciences in 1800, wherein two leading experts in French engineering commented upon the preferability of the Delorme system; and, while they pointed to the need for further trials, they did not identify any fatal flaws in Peale's designs. See Mr. Hart's article, " 'To encrease the Comforts of Life': Charles Willson Peale and the Mechanical Arts," *The Pennsylvania Magazine of History and Biography*, July, 1986, pp. 323–357, especially pp. 348–355.

4. For a remarkably complete and interesting account of the first "Permanent" Bridge over the Schuylkill at Market Street in Philadelphia, see the 84-page published report entitled, "A Statistical Account of the Schuylkill Permanent Bridge, Communicated to the Philadelphia Society of Agriculture, 1806," (Philadelphia, 1807), which was also published as an addition to the *Memoirs of the Philadelphia Society for Promoting Agriculture. . .*, vol. I (Philadelphia, 1808). This includes an annotated plan and partially cut-away elevation engraved by the very skilled Alexander Lawson. Also see Fred Perry Powers, "The Historic Bridges of Philadelphia," *Philadelphia History Consisting of Papers Read before the City History Society of Philadelphia. . .* (Philadelphia, 1917). See also J. Thomas Scharf and Thompson Westcott. *History of Philadelphia* (Philadelphia, 1884, III, 2140–42).

5. See Fig. 11, this book, for some structural details published by John Weale, ed., *Bridges in Theory, Practice, and Architecture*, vol. 2 (London, 1843, plate 35). Also see Theordore Cooper, "American Railroad Bridges," *Transactions* of the American Society of Civil Engineers (vol. XXI, no. 418, July, 1889, plate IV) for the same details, apparently derived from Weale's publication.

6. Some statistical information about the bridge, including the capitalization, debts, dividends, and the average tolls, was published as part of a 23-item questionnaire, the statewide results of which are in "Documents, Accompanying the Report of the Committee, on Roads, Bridges and Inland Navigation, Read in the Senate of Pennsylvania, on the 23d of March, 1822" (Harrisburg). This was an "appendix" to the *Report on*

Roads, Bridges and Canals, read in the Senate, March 23, 1822. The writer is indebted to Willman Spawn for bringing this very informative document to his attention.

7. This statement is from the "Statistical Account. . ." *supra*, n. 4. In that same source, Timothy Palmer is claimed "to have declared" that he would not attempt again a span as large as the 244-ft bridge over the Piscataqua.

8. Although the 1760's Wettingen Bridge (and its European contemporaries) would seem to have been known to the American bridge-building interests, it is difficult to know how such knowledge was transmitted to America except by the various books on travel and the several English encyclopedias that were marketed in the United States and/or reprinted with modifications for American consumption, such as Dobson's encyclopedia, which was published in Philadelphia in the 1790s, or the American edition of the *British Encyclopedia*, or *Dictionary of Arts and Sciences*, or Thomas Pope's *Treatise on Bridge Architecture* (New York, 1811), which, though later, cites several sources for his information about European bridges, such as the multivolume *Wonders of Nature and Art, Repertory of Arts*, and Rees's *New Cyclopaedia*, together with other works on travel and geography. It is difficult to pinpoint the various sources of information about bridge technology, especially when it comes to structural details such as laminated arch construction. Several possible works are cited by Professor J.G. James in his paper, "The Evolution of Wooden Bridge Trusses to 1850," for the Institute of Wood Science: Timber Engineering Group and Institution of Structural Engineers: History Group, London, 1982, pp. 7–8, and n. 28–36. Of course, such works as Pope's *Treatise* or Weibeking's 1810 Bavarian treatise provided some rather specific information about large-scale laminated wooden arches bent to the curve, but they were too late to influence either Peale or Burr, nor did either of them reflect any of the much earlier theories of Delorme. For an excellent study about Delorme's influence on dome construction in America, see the thesis entitled "In Delorme's Manner. . ." by Douglas J. Harnsberger, University of Virginia, 1981. Also see the dissertation by George Michael Danko, entitled "The Evolution of the Simple Truss Bridge 1790 to 1850: From Empiricism to Scientific Construction," University of Pennsylvania, 1979. Despite the many possible sources and studies previously mentioned, it remains something of a puzzle as to the influences upon the earliest American developers of laminated arch construction. Of course, it is conceivable that there was some original thinking on this side of the Atlantic.

9. For a general discussion of Burr's career, including mention of the McCall's Ferry bridge, see the books by R.S. Allen (Note 1). Also see *Documents, Accompanying the Report of the Committee. . .* (Note 6), pp. 172–173. It is this official state report, based on questions submitted to the President and Managers of all the bridge companies in Pennsylvania, that stated that the McCall's Ferry bridge, though started in 1812, was not "completed so as to be passable" until December of 1817 and was destroyed by ice on March 3, 1818. For a more detailed discussion of Burr's experiments with various structural forms in the first two decades of the nineteenth century, see the excellent article by Emory L. Kemp and John Hall, "Case Study of Burr Truss Covered Bridge," *Engineering Issues—Journal of Professional Activities*, ASCE, 10 (E13), July, 1975, pp. 391–412.

10. Burr's Trenton Bridge was admired and discussed in several contemporary engineering works, including Thomas Pope's, *A Treatise on Bridge Architecture. . .* (New York, 1811), pp. 129–138. The single most informative source of information about the bridge is the print from a drawing by C.A. Busby (see Note 11). See also David Stevenson's, *Sketch of the Civil Engineering of North America* (London, 1838), pp. 225–227, and Plate VII; John Millington's, *Elements of Civil Engineering. . .* (Philadelphia and Richmond, 1839), pp. 574–575, and Fig. 248 (which gives a very imperfect impression of the bridge); and Herman Haupt's, *General Theory of Bridge Construction. . .* (New York, 1851), pp. 242–243, and Plate 9.

11. In addition to the treatises mentioned in Notes 3 and 10, especially see the English aquatints by Dubourg from drawings by C.A. Busby, architect, of the "Permanent" Bridge (London, 1823), a similarly formatted print of a Busby drawing (perhaps part of a series) of the "Upper Schuylkill Bridge [Colossus] at Philadelphia from the S.E.," and another very impressive engraving of a Busby drawing of the Trenton Bridge claimed to have been made from measurements made at the site in 1819 (see Fig. 12). It should be noted, however, that the view of the "Colossus" contains some very serious mistakes, and it seems impossible that it represents actual measurements. This raises unanswerable questions about the accuracy of Busby's drawing of the "Permanent" Bridge.

12. The historical outline in this book is largely based upon the Lancaster-Schuylkill Bridge Company Records, Historical Society of Pennsylvania, plus the *Minute Book, 1811–1834*, Thomas A. Biddle Co. Business Books (doc. no. 47), Historical Society of Pennsylvania. These two collections are cited hereafter as LS Br.Co. Records, and *Minute Book*, or *Minutes*, by date. For this writer's earlier historical account of the "Colossus," see *Material Culture of the Wooden Age*, edited by Brooke Hindle, Sleepy Hollow Press, Tarrytown, NY, 1981, pp. 159–183.

13. See the hand-corrected printed draft by William Greer, printer, LS Br. Co. Records, January 30, 1811.

14. Governor Simon Snyder signed the authorizing act on March 28, 1811, *Pennsylvania Archives*, Ninth Series, v.4, 2961. After the required shares were subscribed, the Act of Incorporation was signed on June 10, 1811, *Pennsylvania Archives*, Ninth Series, v.4, 2982; LS Br. Co. Records, June 10, 1811; and *Minute Book*, July 25, 1811.

15. LS Br. Co. Records, June 10, 1811.

16. *Minute Book*, September 24, 1811, and *Poulson's American Daily Advertiser*, Thurs., September 26, 1811.

17. For Mr. Pope's rather detailed proposal (on his letterhead), see LS Br. Co. Records, October 24, 1811.

18. For Mills' proposal to the Board of Managers, see the LS Br. Co. Records, October 26, 1811.

19. *Minute Book*, November 14, 1811. For Lewis Wernwag's "proposition," see the LS Br. Co. Records, November 14, 1811.

20. LS Br. Co. Records, December 5, 1811.

21. The *Minute Book*, December 6, 1811, indicates that a committee was appointed "to give general instructions to Lewis Wernwag to whom the Managers have contracted to build the Bridge," but the resolution to build on "Mr. Wernwag's plan" of a single arch, rather than two arches as had been resolved on November 14, appears in the *Minute Book*, December 20, 1811.

22. For the full inscription on the copper plate which was placed on the cornerstone, see J. Thomas Scharf and Thompson Westcott, *History of Philadelphia, 1609-1884* (Philadelphia, 1884), vol. 3, p. 2144. On January 20, 1812, Wernwag began extensive travels "whilst employ'd in geting [sic] Timber for the Upper ferry Shuylkiln [sic] Bridge Company." See the Simon Gratz Collection, Upper Ferry Schuylkill Bridge Co., Case 14, Box 28, Historical Society of Pennsylvania. The *Minute Book*, February 1, 1812, noted that Wernwag "had taken the lot of ground southward of Mr. Sheridan's property. . .to commence from the 25th day of March next."

23. *Minute Book*, April 30, 1812. For additional information about the Eagle Furnace, see the *Census Directory for 1811*, p. 477; and Charles E. Peterson's article, "Morris, Foxall and the Eagle Works: A Pioneer Steam Engine Boring Cannon," *Canal History and Technology Proceedings*, vol. VII, March, 1988, pp. 207-235. Although the "Estimate of Iron for the upper Ferry Schuylkill Bridge" is probably by Wernwag, it is unsigned and undated; but it is clearly an early estimate because it is based upon five arches, changed to three arches before July, 1812. See "estimates" file, LS Br.Co. Records, n.d.

24. *Minute Book*, May 14, 1812.

25. Earlier, Pope had inquired as to "what Sum as a security (vested in so many Shares) would be an inducement to Your final adoption & erection of a Bridge on My Plan . . .", LS Br.Co. Records, April 4, 1812. Pope's advertisement in the *Aurora* ran from June 12 to June 16.

26. *Minute Book*, June 23, 1812.

27. *Minute Book*, July 2, 1812.

28. *Minute Book*, September 24, 1812. For references to Wernwag's purchasing of an interest in (and management of) the Phoenix Nail Works, see the Thompson Collection, especially document number 44 (various pages), Historical Society of Pennsylvania.

29. *Minute Book*, October 15, 22, 29, November 5, 12, and December 10, 1812.

30. *Minute Book*, December 31, 1812.

31. *Minute Book*, January 7, 1813; and *Poulson's American Daily Advertiser*, January 12, 1813. See also John Wernwag to Samuel L. Smedley, Harpers Ferry, W.Va., August 27, 1874, "Lewis Wernwag, the Bridge Builder," *Engineering News and American Contract Journal* (New York, August 15, 1885), XIV, pp. 98-99.

32. *Minute Book*, February 25 and March 13, 26, 1813; and also see the "rough copy" of the agreement between Robert Mills and the bridge company (apparently in Mills' handwriting), undated, in the LS Br.Co. Records. On April 16, the President was authorized to execute a contract with Mills and to advance Mills $500 on his contract, *Minute Book*, that date.

33. Greville Bathe and Dorothy Bathe, *Oliver Evans. . .* (Philadelphia, 1935), pp. 194-195. Only two weeks earlier, on May 19, Oliver Evans, in a letter to his son George Evans, had rendered a rather poetic engineer's admiration for Wernwag's feat of bridging, ". . . I took a ride this morning to see the upper ferry Bridge on Schuylkill built by Lewis Warnwag [sic]. I think it the Masterpiece of Architecture I ever saw probably the world cannot produce its equal there is not a Mortice or tenant [sic] in the whole arch of 340 feet. It has a most Majestic appearance." Also see the *Minute Book*, July 22, 1813, for reference that Mills "be authorized to examine into the cause of the defect in the Bridge in such way & Manner as he may think proper & Make report at next meeting."

34. On July 1, Wernwag was authorized ". . . to put a Chain along the upper Rib from King post to King post in such manner as he shall think proper Provided that if the Chain should in the opinion of the Board not answer their expectations he is to pay the expense." *Minute Book*, July 1, 1813. On July 8, the *Minutes* note that Wernwag entered into a "rough contract" with the Board "on the subject of removing the defect in the Bridge," but there was no indication about its details. The Statement "exhibited" at the Stockholders meeting on July 31 was entered into the *Minute Book*, August 12, 1813, a large portion of which is included in this book, see the section entitled "The Abutments and the Structural 'Defect'." On August 19, an order was given to put in the "wooden braces between the King posts and a sufficient number of Iron chains under the direction of Mr Johnson." *Minute Book*, August 19, 1813.

35. *Minute Book*, September 23, 1813. For references to Joseph Johnson's involvement with the New Hope as well as the Monongahela Bridge in Pittsburgh, see the *Proceedings* of the Engineer's Club of Philadelphia, vol. III, no. 3 (December, 1882), pp. 179-182; and Herbert Du Puy, "A Brief History of the Monongahela Bridge, Pittsburgh, Pa." *Pennsylvania Magazine of History and Biography*, vol. 30, pp. 188-198.

36. *Minute Book*, November, 1813.

37. LS Br.Co. Records, December 23, 1813.

38. *Minute Book*, January 6, 1814.

39. "Report of the Managers of the Lancaster and Schuylkill Bridge Company to the Stockholders," March 3, 1814 (Philadelphia, 1814). For reference to the "handsome" drawing by Strickland, and the resolution to have it engraved by Kneass, see the *Minute Book*, March 24, 1814. For the order to print 180 copies, 20 others "Handsomely colour'd" see the *Minute Book*, April 16, 1814. For the discussion regarding the wording to accompany the "official" view of the bridge, see the draft of the *Minutes* of the Managers in the LS Br.Co. Records, April 30, 1814. Also in the *Minute Book* for April 30, 1814, is the resolution that the words "general finish by Robert Mills" be added "to the writing under the plate of the Bridge, now in the hands of Mr Kneass."

40. See the *Minute Book*, June 8, 1814; August 19, 1814; January 27, 1815; and LS Br.Co. Records, January 27, 1815. For a detailed letter from Mills to the Managers

regarding work ". . .I consider ought to be done to your Bridge to secure it not only from the effects of the wind but weather, and which certainly comes within the engagements of my contract with you," see the LS Br.Co. Records, January 28, 1815. Among the Robert Mills Papers at the Library of Congress is his 1816 manuscript *Memorandum Book*, p.68, which indicates that, out of his contract for $4520, he had been paid $3900, leaving a balance of $936. As late as November 4, 1816, there is a reference in the *Minute Book* that Mills' claim for completing the bridge was being referred to a committee of arbitrators, but there is no mention of the matter being finally settled.

41. For references to the storm damage, Strickland's estimates, and the necessity to pledge the tolls to pay for the repairs, see the *Minute Book*, September 5, 6, 12, 18, 1821.

42. See "Documents, Accompanying the Report of the Committee. . ." (Note 6), pp. 167–168, which outlines the financial condition of the company and includes a reference to the costs for reroofing the bridge as a result of the storm damage.

43. *Minute Book*, November 6, 13, 1828; May 7, 1829; and June 10, 1829.

44. For minutes of the "special Meeting" on May 9 and again on August 15, 1831, see the Claude W. Unger Collection, Box 60-F, Historical Society of Pennsylvania. Also see the *Minute Book* for those dates, as well as for May 8, 1832, and the LS Br.Co. Records, 1832, for reference to bills for scaffolding and painting.

45. For a newspaper story headlined "FAIR MOUNT BRIDGE DESTROYED," see *Poulson's American Daily Advertiser*, September 3, 1838. There was an advertisement in the same newspaper on September 5–13 offering a reward "for discovery of person or persons, etc who aided, assisted etc in destruction of FAIR MOUNT BRIDGE, paid upon conviction." Note that the common name of the bridge had changed by the 1830s.

46. Ellet's bridge was finished January 2, 1842, and there are numerous sources about its design and construction, but they are not referenced here.

47. *Proceedings* of the Engineer's Club of Philadelphia, vol. III, no. 3 (December, 1882), pp. 179–182.

48. A portion of the text on the broadside entitled "WERNWAG'S BRIDGES," drawn and engraved by E.G. Gridley (Fig. 16).

49. The "experimental" removal of a rib during the construction of the New Hope Bridge was mentioned in the official Report, see Note 47.

50. John Millington, *Elements of Civil Engineering. . .*, (Philadelphia and Richmond, 1839), p. 569.

51. See the "Report of the Managers of the Lancaster and Schuylkill Bridge Company to the Stockholders, March 3, 1814," Philadelphia, 1814, pp. 14–15.

52. The reference to the number of cattle "permitted to pass the bridge at a time" is from the results of a questionnaire published as an appendix to "Report on Roads, Bridges and Canals, read in the Senate [Pennsylvania], March 23, 1822. Mr. Raguet, Chairman." The appendix is entitled "Documents, Accompanying the Report of the Committee, on Roads, Bridges and Inland Navigation, read in the Senate of Pennsylvania,

on the 23d of March, 1822." C. Mowry, Printer, Harrisburg.

53. This statement, attributed to Timothy Palmer, was given in *A Statistical Account of the Schuylkill Permanent Bridge, Communicated to the Philadelphia Society of Agriculture, 1806*, (Philadelphia, 1807), p. 72.

54. This information about the Harrisburg, Northumberland, and Columbia bridges was included in the "Documents," cited in Note 52, pp. 173–180.

55. See the "PUBLIC NOTICE" in *Poulson's American Daily Advertiser*, Thursday, 26 September 1811.

56. *Minute Book*, December 20, 1811.

57. For a general reference to researches and published works on the subject of theoretical and practical problem solving available to engineers, see "Strength of Materials in the Eighteenth Century," Chapter III, by Stephen P. Timoshenko, in *History of Strength of Materials. . .* (New York, 1953). For a contemporary display of early nineteenth-century knowledge about the strength of timbers and other materials (readily accessible to Wernwag), see Thomas Pope, *A Treatise on Bridge Architecture* (New York, 1811), pp. 247–277. Pope mentions the various "experiments" carried out in the eighteenth century, especially as to the compressive strength of wood. For example, he cites experiments (on pages 274–275) where a piece of wood 12 in. by 12 in. could support on its end grain 264,000 pounds, which would be a force of 1833 pounds per square inch. Whether these kinds of figures were useful to Lewis Wernwag is impossible to ascertain.

58. *Minute Book*, July 2, 1812.

59. For reference to Robert Smith's plan and model, see *Pennsylvania Archives*, Eighth Series, VII, 6335–36. In 1786, John Sellers produced a model for a bridge to be erected over the Schuylkill, before the Society of Agriculture, at Carpenters Hall. See *Extracts from the Diary of Jacob Hiltzheimer* (Philadelphia, 1893), p. 103. In November, 1790, there is mention of several bridge models on public view at the State House (Independence Hall); see the *Pennsylvania Packet*, November 2, 1790, p. 3. In 1795, there is mention of a model of Palmer's bridge over the Piscataqua; see the *Minute Book*, Library Company of Philadelphia, vol. IV, p. 18, for February 5, 1795. For mention of eighteenth-century structural models used abroad (and in Philadelphia) for the design of iron bridges, see the article by Emory L. Kemp, "Thomas Paine and his 'Pontifical Matters'" in the *Transactions* of the Newcomen Society for the Study of the History of Engineering and Technology, vol. 49 (1977–1978), pp. 21–40.

60. See *The Aurora*, February 20, 1812, p. 3.

61. D.H. Mahan, *A Treatise on Civil Engineering*. Second edition (New York, 1875), pp. 274–276.

62. See *The Aurora*, June 12, 1812, p. 3.

63. John Wernwag (son of Lewis) commented long after the fact (in an 1885 article, see Note 31) that the piles "were driven to the rock, and at the front of the abutment were 30 feet long, and at the end of the platform only about 15 feet. You can therefore judge what an inclined plane the abutment rested on." While an inclined plane under the piling could explain why there was some movement in the abutment (supported as it was by piling), this explanation doesn't seem to be borne out by the actual subterranean bedrock condi-

tions. Modern river borings taken for bridge construction in the 1960s reveal that the bedrock under the western bank of the river has a slight incline, but nowhere near as much as was suggested by John Wernwag. However, this writer believes that the present Spring Garden Street Bridge western abutment is slightly south of the location of the western abutment for the "Colossus;" and thus it is possible that the bedrock was slightly more inclined on the "Colossus" alignment, see Drawings B6105-CB-1, B6105-CB-2, B6105-2A, and B6105-5A, prepared in 1964 by the City of Philadelphia, Department of Streets, Bureau of Surveys and Design, Bridge Division. The writer is indebted to Nicholas L. Gianopulos, P.E., Keast and Hood Co., Philadelphia; Dave Wismer, Chief of Building Section, Philadelphia Licenses and Inspection; and Jim Leasure, Chief Bridge Engineer for Philadelphia, who made it possible for the writer to obtain copies of these cited drawings, which were utilized (in part) for the preparation of the writer's drawing, Fig. 23.

64. Greville and Dorothy Bathe, *Oliver Evans. . .* (Philadelphia, 1935), p. 194. Entry dated 19 May 1813.

65. *Ibid.,* pp. 194–195. Entry dated 5 June 1813.

66. *Minute Book,* August 12, 1813.

67. Wernwag's several attempts to remedy the defects, including the removal of the earthen fill within the abutment and replacement with stone (which is mentioned in the Managers report to the Stockholders, see Note 66), and his work to "cut away the stone behind the heels of the two ribs that were high in the center and let them back to their proper position" (which was *not* mentioned in the report) was related by John Wernwag to Samuel Smedley, long after the fact, in a letter dated August 27, 1874, and which was printed in an article cited earlier, Note 31.

68. For possible "laminated" arch influences upon Wernwag, see the illustrations in the more recent book by Werner Blaser, *Wooden Bridges in Switzerland* (Basel, 1982), pp. 58–80.

69. The writer is indebted to a number of individuals who, over the years, have provided assistance and advice in a variety of ways: Willman Spawn of Philadelphia, for calling attention to a number of rare and relevant items among the collections at the American Philosophical Society; Richard Sanders Allen, for sharing information about Wernwag as far back as 1958, which is when this study first got underway; Robert M. Vogel, for similarly being very responsive to inquiries and sharing files and researches from the Smithsonian Institution, to say nothing of his own knowledge and judgement on several matters relating to the "Colossus," beginning in 1961; officials at the Atwater Kent Museum, the Franklin Institute, and the Free Library (all in Philadelphia) for courtesies in making it possible to take photos of museum objects in their collections and/or providing copy negatives for my use, mostly in the early 1960s; Charles E. Peterson, FAIA, for several useful items related to Wernwag and his contemporaries; Eric DeLony, *Historic American Engineering Record* (HAER), National Park Service, for his knowledge about bridge construction technology and for being a ready sounding board for ideas in the later evolving stages of this study when it shifted from the "Colossus" to a context of "laminated" arches; Richard Anderson (also of *HAER*) and Sharon Park, AIA, National Park Service, for especially good advice that helped the writer in the preparation of the several drawings prepared for this study; the writer's son, Allan L. Nelson of Middletown, Delaware, for his ready help in making one of the drawings; Emogene Bevitt, National Park Service, for very useful editorial advice on a number of aspects during the final writing of the manuscript; and special thanks to Nicholas L. Gianopulos, P.E., Keast and Hood Co., Structural Engineers, Philadelphia, for his generous offer to undertake the GT STRUDL computer analysis of the "Colossus;" and to Jon E. Morrison, P.E., also of that firm, for carrying out the analysis, for sharing his engineering perceptions about the "Colossus," and for writing the "Summary Report" which is included in this book; and to Lisa D. Amthauer, also of that same firm, for preparing a "Force Diagram," which the writer used as a prototype for the preparation of Fig. 22. These contributions from the firm of Keast and Hood Co. provided the writer with a much sharper engineering understanding of the bridge, and they constitute an important aspect of this book. Also, thanks and appreciation are due to Emory L. Kemp, Fellow, ASCE, Professor, College of Arts and Sciences, West Virginia University, for reading this manuscript and for making very constructive suggestions.

Finally, the writer wishes to thank Margaret Thomas Will, formerly with the Technical Preservation Services Division, National Park Service, Washington, D.C., now living in Munich, Germany, who, during the final stages of this report, provided some new information about Lewis Wernwag's origins. Heretofore, it appears that most of our knowledge about Wernwag came from his son, John, who communicated some biographical data about his father in an 1874 letter to Samuel Smedley, which was published in 1885 (see Notes 31, 63, and 67). John's data may have been the basis (at least in part) for the entry in *Appleton's Cyclopaedia of American Biography,* (New York, 1893), vol. VI, p. 437. The "new" information about Wernwag comes from the multivolume series entitled *Deutsches Geschlechterbuch* (German Lineage Book, Genealogical Handbook of Untitled Families), vol. 34, 1921, pp. 421–447. From this compendium of archival information, it would seem that Lewis Wernwag was born in Reutlingen, not in Riedlingen (John Wernwag's 1874 letter may have been misread by Samuel Smedley, and the published error was repeated by subsequent sources). Lewis' father, Johann Ludwig Werenwag (1729–1791), was a butcher and was a leaseholder of the Alteburger Hof at Reutlingen. "Our" Lewis, the third son, was born in Reutlingen and was named Johann Ludwig Werenwag (after his father), but in America, he went by the name Lewis Wernwag. Following the entry for his name in the aforementioned genealogy is a footnoted reference to an article by Otto Lohr, "Sieben Schwaben in Amerika" (Seven Swabians in America), in *Der Schwabische Bund (The Swabian Union),* vol. 3, December, 1920, pp. 189–194, which identifies "our" Lewis (or Ludwig) as one of the most famous Swabians in North America, especially prominent as a bridge builder and a "self-made-man." This 1920 outline of Lewis' life seems to have been derived from previously published sources and thus is of little value, but the genealogical compendium cited earlier has very precise data about Wernwag's family origins and would provide useful background information for a future study of Wernwag and his multifaceted career in America.

For the present, most of what we know about Wernwag's life and career comes from the 1874 letter (mentioned earlier) from John Wernwag (his son) to Samuel Smedley, a Philadelphia engineer; but since the facts and dates in that letter were based upon John's memory, some of that information has proved difficult to verify. Although there are numerous other reliable sources about Wernwag's activities, much research remains to be done; therefore, the reader is warned that the following chronology is tentative as to the exact dates and the specific nature of Wernwag's role in the activities mentioned. The chronology is presented here to help place Wernwag in context with his contemporaries and to suggest the scope and variety of his activities.

A Brief Chronology of Lewis Wernwag (1769–1843)

4 December 1769, born in Reutlingen, Württemburg, Germany. There is a family tradition that he was secreted in the mountains by a shepherd (to avoid being conscripted during the Napoleonic Wars), that his attention was directed to the study of astronomy, natural history, and other scientific subjects, and that he made his way to Amsterdam.

ca. 1786, came to America and settled in Frankford, Philadelphia.

1790 ff., worked as a mill-wright, building materials supplier, and building contractor in and around Philadelphia.

ca. 1809, purchased land containing large quantities of oak and pine, from which he got out the keel for the first frigate built at the Philadelphia Navy Yard.

1810–1812, built two small wooden draw bridges just north of Philadelphia, one over the Neshaminy Creek and one over Frankford Creek.

1810–1811, was the masonry contractor for the "fireproof" vaulted construction of the Burlington County Prison in Mount Holly, N.J., designed by Robert Mills.

1812–1813, designed and built the "Colossus" at Fairmount, in Philadelphia.

28 March 1812, patented his bridge designs, probably including the "Colossus."

1812, bought an interest in the Phoenix Nail Works, Chester County, Pennsylvania, and removed to that place with his family to manage this large cut and wrought nail manufactory, including a rolling and slitting mill. In 1816 he built a new factory there with 42 nail machines at work. He also purchased very large quantities of anthracite coal and lumber, for use in the mill and to sell, and ranged far afield to build bridges, dams, canals, and did survey work, etc. His share of the works, including the iron works and buildings, was sold at Sheriff's Sale in 1820.

1813–1814, built (with Joseph Johnson and Samuel Stackers) the New Hope Delaware Bridge, an important early example of a six-span laminated bow-string truss, each span of which was 175 ft long.

1814–1816, built (with Joseph Johnson and Issac Nathans) the Penn Street Bridge over the Schuylkill at Reading, with three spans of 200 ft each (utilizing the structural principles of the New Hope bridge), and which was embellished with carved sculptured figures of Agriculture and Commerce by William Rush; Wernwag was hailed as "the Pontifex Maximus of these United States."

1815, bought five shares in the Schuylkill Navigation Co.

1816–1818, designed and built under the firm name of Wernwag and Johnson (Joseph Johnson) the enormous eight-span wooden bridge (1500 ft long) over the Monongahela at Pittsburgh at a cost of $110,000.

1817, advertised to lumbermen on the Schuylkill that he would buy no more lumber but would purchase rafts of fence rails loaded with coal.

1817, contracted to build a section of the Schuylkill Canal in Schuylkill County, including the Mount Carbon Dam.

1817, designed the three-arched bridge (completed by Isaac Nathans) on the New Hope plan at Schuylkill Falls, near Philadelphia.

1817–1819, designed and built the Wilkes-Barre Bridge over the Susquehanna, patterned after the New Hope Bridge but designed with four spans of 185 ft each.

1817–1819, built (with Joseph Johnson) the long (1122-ft) six-arched bridge over the Allegheny River at Pittsburgh; both Pittsburgh bridges destroyed by fire in 1845.

1818, advertised for sale Schuylkill coal, coal stoves, and sawn hemlock rails.

1818, built bridge at Jones Falls in Baltimore.

ca. 1818, moved to Conowingo, Md. where he started the multispan bridge across the Susquehanna, established a saw mill, and prepared timbers for the construction of other bridges.

1820, completed the ten-span Conowingo Bridge over the Susquehanna, with two of the spans of "upwards of 210 feet in length," and one at 199 feet.

1820–1821, built bridge over the Brandywine, at Wilmington, Del.

1822, built bridge at Paulings Ford near Phoenixville over the Schuylkill.

1822, was building a bridge at Nashville, Tenn. (mentioned in a law suit).

1822–1823, built a pile and trestle bridge over the Choptank at Cambridge, Md.

1824 ff., moved to Harpers Ferry (then in Virginia, now West Virginia), bought three acres of land on Virginius Island, situated along the Shenandoah River, for $8000, where he built a mill race, a saw mill, a three-story machine shop, and dwelling houses; also built the Wager Bridge, 750 ft total length over the Potomac River (the first bridge at Harpers Ferry), built other buildings including the Methodist Episcopal Church (1825–1828) and the Free Church (1827), and prepared timbers for other bridges that he and his sons built in succeeding years.

1824–1825, built bridge over Goose Creek in Loudoun County, Va.

ca. 1825, built bridge over the Gunpowder River, Md.

1827–1830, involved with his son (L.V. Wernwag) with a bridge at Cambridge, Ohio, over Wills Creek.

1828, built a timber and stone dam (about 10 ft thick and 1800 ft long) across the Potomac River for the Harpers Ferry Armory.

1828, published a 12-page "Memorial" to Congress, "Praying an Extension of his Patent for his Improvement in Bridge Architecture," datelined Harpers Ferry.

1829, patented an improvement on his original bridge patent.

16 November 1831, patented a "self-directing" railroad car and was constructing and selling this "Self Regulating" car that was intended to "run around curves."

1831, built wooden bridge over the Monacacy River, near Frederick, Md., for the Baltimore and Ohio (B & O) Rail Road, with three arches of 110 ft each, from the designs of B.H. Latrobe, Jr.

1831 ff., designed bridge (built by sons Lewis and William) in Indiana over W. Fk. White River.

1832, firm of Wernwag and Sons dissolved, sold half of the interest in their tract on Virginius Island; a new partnership was developed, which specialized in the lumber business.

1833, he and his wife Elizabeth sold the remaining interest in Virginius Island to their son John, and Wernwag retired from the business of the saw mill.

1834, possibly designed the long-lived bridge (built by Josiah Kidwell) on the Northwestern Turnpike over Cheat River near Erwin (now W. Va.).

1833–1834, built a pivot bridge over the Chesapeake and Ohio (C & O) Canal at Point of Rocks, Md.

1834–1835, possibly built bridge at Romney (now W. Va.) on the Northwestern Turnpike over the south branch of the Potomac.

1835–1836, built (but did not design) the very large wooden railroad bridge which crossed both the Potomac River and the C & O Canal, at Harpers Ferry.

1836, built a small bridge over the Shenandoah Outlet Lock of the C & O Canal.

1837, had patented and was manufacturing (at his plant on Virginius Island) and selling a "Patent Cutting Box," for cutting provender for stock.

1838, probably designed a 240-ft clear span bridge (built by son L.V. Wernwag) at Camp Nelson, Ky., over the Kentucky River, which in 1927 was subjected to a stress analysis, published in *Engineering News-Record*, with a maximum unit compression in the arch of 1,080 pounds per square inch.

12 August 1843, died at Harpers Ferry, at the age of 73.

INDEX

( ) Denotes Figure Number
Index includes persons, places and projects mentioned in the text, including the Wernwag Chronology